Going for Growth

Going for Growth

Technological Innovation in
Manufacturing Industries

R. C. PARKER
Ashridge Management College

JOHN WILEY & SONS
Chichester · New York · Brisbane · Toronto · Singapore

131310

Library of Congress Cataloging in Publication Data:

Parker, R. C. (R. Charles)
  Going for growth.
  Includes bibliographies and index.
  1. Technological innovations—Great Britain.   2. New products.   3. Industrial management—Great Britain.
  I. Title.
  HD45.P333 1985   338'06   84-21902
  ISBN 0 471 90633 6

British Library Cataloguing in Publication Data:

Parker, R. C.
  Going for growth: Technological innovation in manufacturing industries
  1. Corporations—Growth   2. New products
  I. Title
  338.7'4   HD2746

  ISBN 0 471 90633 6

Printed in Great Britain by
St Edmundsbury Press, Bury St Edmunds, Suffolk

To my daughter - Cheryl Elizabeth

Acknowledgements

The author is indebted to the Department of Industry, who financed a project 'Innovation in Smaller Companies' from September 1980 until May 1982; this provided the material incorporated in Chapter 7. In June 1982 a further study on the growth of manufacturing businesses through technical innovation was undertaken, and the author is extremely grateful to the following bodies and companies whose generous funding made this work possible: Blue Circle Industries plc, British Technology Group, Commercial Union Assurance Company plc, Grand Metropolitan plc, National Westminster Bank plc, The Prudential Assurance Company Limited, Steetley plc, and Wiggins Teape Research and Development Limited.

It is a pleasure to acknowledge the support and encouragement which I have received throughout from Mr Philip Sadler, Principal of Ashridge Management College. He generously took time to read the draft manuscript, and his many suggestions have proved invaluable.

I also wish to thank my secretary, Elizabeth Pitcher, for cheerfully and patiently dealing with what must have seemed endless revisions; and Mr Ken Walker for the skilful manner in which he converted very rough sketches into illustrations. Grateful acknowledgement is also due to Mr V. Nolan for writing Appendix I in Chapter 7. I am particularly indebted to Mrs K. Paterson for her kindness in compiling the index. Finally, thanks are due to my wife's supportive interest, for, without her perception clarity would often have been sacrificed for brevity.

We are grateful to the following for permission to reproduce copyright material:
Basil Blackwell
Figures 4.2 and 4.3, from Allen (1970), *R & D Management*, **1**.

Figures 8.1 and 8.3, and adaptations of Figures 8.2 and 8.4, from Parker and Sabberwal (1971), *R & D Management*, **1**.

Figure 4.1 adapted from Le Coadic (1974), *R & D Management*, **4**.

Figure 7.8, from Carson (1974), *R & D Management*, **5**.

Figure 2.1 adapted from Jermakowicz (1978) *R & D Management*, **8**.

Figures 3.6 and 3.7, adapted from Wissema (1982), *R & D Management*, **12**.

Figure 6.4, adapted from Wilkinson (1983), *R & D Management*, **13**.

Tables 9.1 and 9.2, adapted from Parker (1983), *R & D Management*, **13**.

Figure 4.6, from Saren (1984), *R & D Management*, **14**. British Institute of Management Foundation

Figure 2.4, from Parker (1980), *Guidelines for Product Innovation*. J.P. Coleman and the Institution of Electrical Engineers

Figure 1.1, from Coleman (1970), *Electronics and Power. Council of the Institution of Mechanical Engineers*

Figures 3.1, 3.2, 3.4 and 3.5, from Parker (1970–71), Proceedings of the Institute of Mechanical Engineers, **185**.

Figure 9.5, from Parker (1974), *Conference Publication*, **C140**.

Figures 8.5 and 8.6, from Parker (1975), *Automobile Proceedings*, **189**.

Figure 7.7, from Parker (1980), *Chartered Mechanical Engineer*, **27**.

Figure 6.1 from Parker (1982), *Chartered Mechanical Engineer*, **29**.

Figure 4.7 from Neale (1984), *Chartered Mechanical Engineer*, **31**. EC Commission

Figure 6.2 from Piatier (1980), *Barriers to Innovation in European Community Countries*, No. XIII(81)04. Engineering Magazine

Tables 9.4 and 9.5 from McRae (1979). Harvard Business Review

Figure 2.2, from Greiner (1972).

Figure 3.12, from Chambers Mullick, and Smith (1971). HEURITA, Zurich

Figure 7.9, from Pugh (1981), *International Conference on Engineering Design*. Industrial Management Center, Inc.

Figures 3.3 and 3.8 from British (1978), *Practical Technology Forecasting*. Institute of Measurement and Control
Figures 3.9 and 3.10, from Lunt (1975), *Measurement and Control*, **8**. Institution of Production Engineers
Figure 1.2, from Corlett (1983), *Production Engineer*. McKinsey & Co. Inc.,
Figure 6.5, from Allen (1978), *McKinsey Quarterly*. Macmillan
Figure 5.4 9.1, adapted from Kay (1982), *The Evolving Firm*. MCB University Press Ltd.
Table 9.3 from O'Leary (1978), *Management Decision*, **16**. MIT Press
Figures 4.4 and 4.5, from Allen (1978), *Managing the Flow of Technology*. National Westminster Bank plc
Figure 9.1, adapted from Jones (1978), 'Our manufacturing industry—The Missing £100,000 million', *National Westminster Bank Quarterly Review*, May 1978. NMB Bank, Amsterdam
Table 2.1, from Velu (1980), 'Entrepreneurial characteristics', *European Foundation for Management Development's 10th European Seminar on Small Businesses*. North-Holland Publishing Co.
Figure 6.3, adapted from Holt (1978), *Research Policy*, **7**. Oil and Colour Chemists' Association
Figure 3.11, from Smith (1981), *Journal of the Oil and Colour Chemists' Association*, **64**. Pergamon Press Ltd.
Figure 9.3, from Cox (1977), *Long Range Planning*, **10**.
Figure 7.6, from Robinson, Hichens and Wade (1978), *Long Range Planning*, **3**. Frederick Polhill (Marketing Consultants)
Figure 7.4, from Wearden (1981), *Industrial Marketing Digest*, **6**.

Contents

Foreword

In Charles Parker's previous book, *The Management of Innovation* (John Wiley 1982) he presented eight case studies of innovation in British manufacturing industry and distilled from these some ideas about ways in which the management of innovation could be improved. In this new work he sets the important subject of innovation and new product development in a wider context, looking first at its relationship to economic growth and national prosperity, then at its role in the evolution of the business. He then turns to key issues and processes involved in the innovation process itself—creating the climate for growth and innovation; forecasting changes in the environment; the communication processes involved both within the company and externally; the search for ideas and the actual development of new products; the relationship of new product development to strategic planning; the management of R & D, sources of finance and the role of government and government agencies.

All this adds up to a comprehensive treatment of an aspect of management which is vital to industrial survival—not only at corporate level but also nationally. It is a work which should be of considerable use both to the practising manager and the student of management. It represents an approach which fits very well with the philosophy of Ashridge Management College in that it is rigorous without being academic or arcane and is practical and relevant without being simplistic. Charles Parker represents a type of management researcher who is all too rare in our society—someone who has turned to systematic research in the management field following a distinguished career as a research scientist and director of research in industry. To approach the study of management from such a background is to be armed with a mixture of practical experience and intellectual capital, which, when brought to bear on the issues and processes involved in something as complex as new product development can analyse the subject without rendering it obscure and can illuminate it without being superficial.

I am very pleased that it has been possible for Dr Parker to carry on his valuable work as a Senior Research Fellow of Ashridge Management College. Ashridge exists to contribute to the improvement of the practice of management and seeks to do this both by teaching and research. This, the latest of its research outputs, constitutes a major contribution to our knowledge and understanding of the management process and to our ability to improve on past performance standards.

PHILIP SADLER
Principal
Ashridge Management College

Introduction

My interest in innovation arose from spending 29 years in the research and development division of a manufacturing company which enjoyed steady growth despite ever-increasing world competition. The company, which was a supplier to most engineering manufacturers, had to adjust to contraction of mature industries and recognize those technical advances which were likely to form the bases of future growth. It was, additionally, necessary to replace evolutionary product development by a more radical approach.

These experiences gave a fascinating insight into the crucial contributions of product innovation to company growth, and when it was suggested that studies on this topic should be undertaken within firms, with Ashridge Management College as the academic base, the opportunity was enthusiastically accepted.

The initial management studies comprised two projects: 'The Management of Innovation', and 'Innovation in Smaller Companies'. The purpose of the first was to determine those factors which either facilitated or hindered innovation, and its objective was defined as 'to stimulate increased innovative developments within manufacturing industries by carrying out practical case studies, with the cooperation of selected companies, of the good and bad factors controlling innovation within the organization.' The second had a similar aim which was to design an advisory service that would help small companies to innovate. The sponsors of both studies emphasized their preference for information which could be of practical help, and this made it easier to avoid the temptation of carrying out a purely descriptive or analytical exercise, and to adopt, instead, an approach which was designed to yield prescriptive outcomes.

In the first project, discussions were held with 43 manufacturers, from whom eight were chosen for extended studies. The first

first consideration when selecting companies in which to work was to ensure that, between them, they encompassed all stages of the innovative process, since the duration of the work was less than the time which normally elapses between an idea for a new product and its launch on to the market. Other criteria were that a business had to be United Kingdom owned, and had either to be independent or an autonomous subsidiary of a group. The final selection was made in the expectation that the projects would be hosted by companies keen to innovate, willing to cooperate, and would represent a wide range of sizes, types of ownership, products, structure, organization, and manufacturing processes.

In four of the companies selected for the first study help was given in developing an organizational climate in which new ideas could be generated and encouraged, and in the other four the study was mainly observational. Regular visits to the companies were made over periods which extended to three years. Halfway through the first project, guidelines for product innovation were drawn up and published (1) and towards the end a concept was introduced which enabled an innovative company to be represented by a model based on six designated levels of technology. This model served to indicate which of the guidelines were relevant to a particular level of technology, and facilitated the formulation of optimum strategies directed towards increasing turnover, profit, and growth (2).

The second project was built upon experience gained from the first. Three criteria were used for selecting firms in which to work. An industrial classification was only judged suitable when small and medium sized enterprises contributed over 25 per cent of total numbers employed, whilst the number employed in the five largest firms had to constitute less than 40 per cent. The number of employees was required to range from 50 to a 1000. Discussions were held with the chief executives of 22 firms, and nine were selected for an extended study.

Data from the first study showed that of the 43 companies visited 22 were judged to be in need of help in their search for new products, and, of these, only ten possessed the necessary resources for satisfactory product development.

The second study gave a picture that was surprisingly similar in view of the studies' small populations. Of the 22 companies visited just over a quarter said that they had an urgent need for new products, possessed the necessary resources, but had little or no experience in either initiating or managing new products. They, nevertheless, invariably responded with enthusiasm once it was demonstrated that new opportunities could be gained from a structured search for new products which could be developed with existing resources. It is believed to be significant that all companies in this section were subsidiaries of large groups. A further five companies, also subsidiaries, were in dire need of new products but were judged to have attitudes that were too negative for success to be achieved within the time-scale of the project.

The remaining half of the companies with which the second project was concerned were under the control of the original entrepreneur. They were all successful and although the majority were content to remain small, they had the means and ability to expand had they so wished. These enterprises had a number of important features in common which were not shared by the subsidiary companies. The products of these successful companies which were unique, durable, reliable, and well-designed, met an outstanding need, and reflected an intimate knowledge not only of the market but of competitors' activities. Their export performance was impressive: two exported no less than 45 per cent of their output, and three achieved 85 per cent. The calibre of the scientific and technical staff was outstanding. Figure A illustrates the combined observations from both studies.

The negative attitude of the majority of companies in both arms C and D of Figure A reflected the strategy of the parent board. Little attention had been given towards planning long-term growth, and hence their inevitable response to the recession had been the disposal of loss-making subsidiaries, reduction of both manning and working capital, and the contraction of the business to its core activity. When development work was directed towards new products, it was, more often than not, restricted to the replacement of obsolete materials with little thought given to new markets. It was a realization of the dangers, inherent in this situation, that gave rise to a third study on the growth of manufacturing businesses, the results of which form the substance of this book.

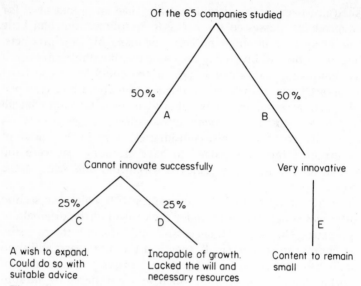

Figure A.   Growth potential based on innovation

If too little emphasis seems to have been placed upon a number of the more erudite theses it is deliberate, for the reason that they were thought unlikely to make a practical contribution to growth in the near future, and would therefore make the book of less immediate interest to practitioners. The need throughout the world is for manufacturing industries to increase prosperity now, and practical observations suggest that a large potential for industrial expansion could be achieved through greater application of existing knowledge. Books intended for practical use by executives and managers should be concise, and, for this reason, the treatment of subjects where scope is particularly broad has been restricted to general principles. A literature reference is provided at the end of each chapter for readers who wish to explore the more complex concepts in greater detail.

REFERENCES

1. Parker, R. C. (1980). *Guidelines for Product Innovation*, British Institute of Management Foundation, London.

2. Parker, R. C. (1982). *The Management of Innovation*, John Wiley and Sons, Chichester.

1

Industrial Growth and Prosperity

1.1 SOCIETY'S DEPENDENCE ON INDUSTRY

The elimination of poverty, and the quality of life, depend upon manufacturing industries being able to generate wealth. Part of the wealth will fund defence, education, and social services; part will be spent on goods and will so further encourage industrial activity. The industrial creation of wealth, illustrated in Fig 1.1(1) shows that the basic operation is the addition of useful work to materials in order that they may be sold at a profit.

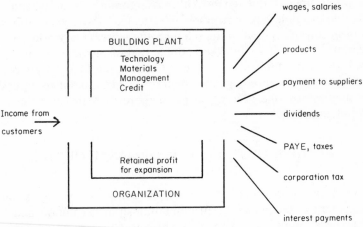

Figure 1.1 The generation of wealth

A plea for a widespread respect and an esteem for industrial activity has been made by Adams(2). He believes that there is no more moral, nor more socially responsible task for a person to take up, or for an older person to continue to pursue, than that of working in industry to produce the necessary goods and services to relieve poverty. No-one can consume that which has not been produced, and no-one can give to anyone else—no matter how charitable they may be—material goods which have not been produced by their own, or someone else's hands. Smith(3), in his paper entitled 'The task of modern industry' contends that without growth of industrial activity countries cannot attain stability, strength, and prosperity.

The wealth creation process depends upon three functions being adequately performed—agriculture, manufacturing, and extraction. The less developed countries are concerned mainly with adding value by extracting minerals or by gathering and preparing crops. Most developed countries direct their efforts towards manufacturing, but a few use considerable resources in the application of advanced technology to large-scale extraction of oil, coal, and other minerals.

In the long term economic growth has enabled the needs of man to be met with considerable success. There are numerous manifestations of our innate skill in overcoming obstacles, and a notable example has been our ability to feed those populations subject to rapid growth rate. Though the increase in world population from 1850 to 1950 was double that of previous centuries, and was again doubled in the next 30 years, the food supply per head is, today, not only greater than it was in 1750 but continues to increase by one percentage point more than the increase in population.

1.2 THE INDUSTRIAL PERFORMANCE OF THE UNITED KINGDOM

The spread of manufacturing industries across the world was an important consequence of Britain's industrial revolution and a major example of the manner in which wealth may become distributed among countries. Its early outstanding achievements were, unfortunately, not sustained and some of the principal reasons for this merit comment because they give an indication of the corrective action which needs to be taken.

Statements about the relative decline of United Kingdom economic performance since the mid-19th century are commonplace, but a particularly instructive and detailed account of this has been given by Carey(4). To take but one example from his analysis, Britain's share of the world's exports in manufactured products decreased from 40 per cent in 1870 to 9 per cent in 1979. Corlett(5) has, more recently, pointed out that during the first quarter of 1983 our balance of trade on manufactured goods became negative for the first time in 200 years (see Fig 1.2).

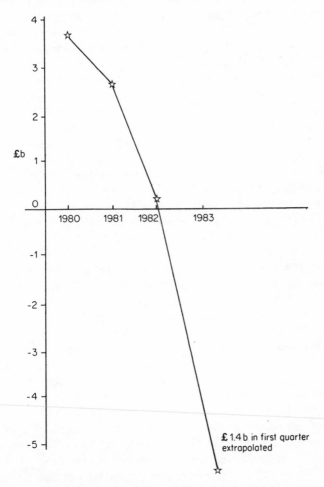

Figure 1.2 The United Kingdom decline in the balance of trade of manufactured goods (from the *Guardian*, 13 May 1983)

Though the beginning of 1984 heralded some promise for industrial renaissance, the European Management Forum(6) rated the United Kingdom as very poor in terms of competitiveness and productivity with a placing, in 1983, of 14th out of the 22 major industrial countries.

The origin of the United Kingdom's successes and subsequent failures was the agricultural and industrial revolution which lasted a hundred years from 1740. One feature which had far reaching consequences was the fact that villages lost many of their craft industries and became only sparsely inhabited, while towns, by contrast, became crowded; the population of England and Wales doubled, and that of Lancashire quadrupled. Trevelyan(7) has argued that this shift of population had the important consequence that whereas countrymen had regarded poverty and oppression as an individual misfortune, for the urban workers it became a group grievance. The disparity of wealth became so great that the social injustices gave rise to conflict which became centred upon class differences. A particularly damaging and long lasting consequence was a lack of common purpose between employer and employee, a process that became accentuated by hardship caused by the 20 year war with Napoleonic France. Indeed, antagonisms of such intensity were to develop that they continue to the present day and still influence current industrial attitudes. As a culture developed which was antipathetic towards industry, our impetus faltered, and neighbouring countries quickly recognized the substantial advantages attainable from industrialization, and began, successfully, to attack our markets.

Barnett(8) has described clearly the particular attitudes associated with later phases of industrialization which have remained as a deterrent to industrial growth. The interesting question is why our European neighbours, many of whom experienced social upheavals of a similar character, were better able to mount a programme of industrial resurgence in the 19th century. This matter has been briefly examined by Parker(9) who has ascribed most of the blame to the parlous state of our education dating back over a hundred years. The early part of the 19th century saw a remarkable downward trend in education. Grammar schools closed down in large numbers and public schools also suffered a temporary decline. The Grammar School Act of 1840 portrayed clearly the prevailing attitude towards technology in that it restricted such foundations to classical education. Indeed, much remains to be done in the field of education before a vigorous growth culture can again be established.

Developing countries, which are seeking to increase wealth through manufacturing activities, will find in the United Kingdom industrial history a rich source of experience. This demonstrates a need for management and the workforce to agree objectives and work in harmony. Education and institutions must adopt an integrative role in forging a supportive relationship between science, technology and society. Governments must accord a higher priority to the creation of wealth than to its distribution, and this will call for equal opportunity rather than egalitarianism, and for an elite comprised of those who demonstrate excellence.

1.3 FUTURE EXPANSION OF WORLD MARKETS

The thesis that jobs could be provided for all those who can and wish to work once manufacturing companies attain the expansion of which they are capable, presupposes a gigantic potential worldwide appetite for consumer goods. It will now be argued that there is evidence to suggest that this will be the case, although it is currently hindered by the failure of governments to cooperate in a manner which would result in economic policies that would equally benefit all nations. The difficulties of so doing are, of course, immense, for not only are there disagreements on economic theories, but the likelihood of success is hindered by interactions between peoples' attitudes and government actions. Once conditions are seen to worsen people aggravate the situation of which they are apprehensive, by falling prey to ignorance, prejudice, and fear. However, there is cause for optimism because international cooperation, already at a high level, will be made easier through technical advances which are now being achieved in communication and travel. For example, response to national disasters is now worldwide and immediate; goods are collected for despatch to stricken areas almost within hours, and to a degree which would not have been possible a decade or so ago. The need is to arouse this level of international cooperation and direct it to important economic problems.

A significant pointer to the inevitability of an eventual resurgence in world markets is the adoption by countries, including those in the Soviet bloc, of economic programmes under the auspicies of the International Monetary Fund (IMF). The vast loans, the rescheduling of debts, and the informal concerted action for mobilizing aid by the IMF, governments, central, and commercial banks, are undoubtedly based upon a conviction that sustainable worldwide growth will eventually permit a repayment of both interest and debts. To those who doubt the ability and will of developing countries to restructure their debts, it is interesting to note that Kaletsky(10) quoted the US Treasury Secretary, Donald Regan, who recalled that the US international debts incurred in 1840 were not repaid until the First World War. As a debtor nation they continued to borrow more abroad and experienced moments of panic, but their creditors stayed with them until their wealth increased and all debts were cleared.

Two encouraging facts should be noticed among the United Kingdom statistics that have been published during the last three decades. The first is that the United Kingdom national income increased continuously in absolute terms after the mid 1950s and grew at an average rate of 2.5 per cent between 1955 and 1974; and the second is that, even so, only a small fraction of full potential was achieved.

The inference is that the United Kingdom, despite its bad comparative performance, could have been capable of a much higher production which would have created a greater demand for consumer goods and this, in turn, would have stimulated imports and exports. It also seems reasonable to assume that the United Kingdom's more successful rivals, with a similar economic structure, namely Belgium, France, the Netherlands, and Italy, could also have increased their rates of growth to a level that would have more closely approached that currently obtained by Japan. There is thus reason to believe that the economic potential of the advanced countries still greatly exceeds their achievement. Carey(4) suggests that, because, until fairly recently, the standard of living in the United Kingdom underwent continuous improvement, industry was not sufficiently motivated to deal with the then difficult problem of low productivity. This reminds us of Gabor's(11) observation that 'man is wonderful in adversity, but weak in comfort, affluence and security'.

The inevitability of a revival in world trade is supported by a number of considerations. Piatier(12) has argued that if short-term and partial analysis, which usually leads to pessimistic predictions, were replaced by a broad view of the economy the indications are that we are in the embryonic phase of a fourth long-wave economic cycle. To appreciate the possible consequences of a new upsurge in technology it is pertinent to recall that Britain, despite her weakening comparative position, was, by reason of her unique position at the hub of the world's economy during the mid-19th century, still able to effect a profound influence upon the international diffusion of industrialization. Distribution of manufactured goods was followed by the export of technology and machinery; this stimulated less developed countries to produce more crops, extract greater quantities of minerals, and begin production of goods. It is heartening to ponder upon the spectacular manner in which means of transport spread to all countries. Cars and commercial vehicles soon became regarded as basic necessities and, indeed, as creators of growth and wealth.

Because technology has now become a part of almost all of our activities it is easy to forget the tremendous contributions that can flow from one scientific discipline. Dörfel(13) has reminded us of the astonishing changes which chemistry has made to fundamental conditions in various walks of life, and to society, since the end of the 19th century. Chemotherapeuticals and antibiotics have reduced the death rate from infectious diseases in Germany from 20 per cent of all deaths in 1927 to 1 per cent in 1976. When in 1798 Malthus made his famous prophesy that mankind was doomed to die of starvation, he foresaw neither that investigations into plant structure would identify the crucial role of phosphorus, potassium and nitrogen, nor that the inadequate source of naturally occurring mineral fertilizers would be augmented through the large-scale production of ammonia by the Haber–Bosch process.

Single discoveries are similarly capable of influencing the economy and society in unpredictable ways. Perhaps the question asked of Michael Faraday 'Of what use is electricity?' may seem naive but less dramatic; yet, nevertheless, significant instances of unforeseen developments are common. One of the important events in the last 30 years was the official announcement by William Schockley's team in 1948 of the discovery of the transistor(14). Few were aware of its potential and in 1952 electronic valves achieved their peak production. By 1956,

however, 164 different transistor types were on the market, and
the researchers were awarded the Nobel Prize. Today there are
as many as 450 000 integrated circuits mounted on a single sili-
con chip, and the industry is worth billions of pounds. Even a
single development can have a significant effect on employment
and wealth. In 1982 videos were estimated to have created 27 000
jobs in Britain(15) in the fields of distribution and cassette manufacture, and at a time when every instrument was imported.

The last two decades have seen a long list of innovations
which include microelectronics, information science, solar power,
robotics, aquaculture, fibre optics, biotechnology, medical diag-
nostic systems and, of course, many whose growth remains hidden. There are thus sound reasons for optimism, and especially
so since, compared with the 18th century there are now many
more countries which are able to act as sources of invention and
diffusion of technology.

The possibility of a favourable outcome from the above scenario is strengthened by the fact that the highest growth rates
have recently been achieved by a number of newly industrializing
countries. For example, in the period 1970–80 the per cent annual growth rate of South Korea, Hong Kong, and Malaysia were
9.5, 9.3 and 7.8 respectively(16). A prediction(17) that by AD
2000 third world manufacturers will account for 20 per cent of
industrial output compared with 7 per cent at present, suggests
that the process will continue. Their export of indigenous raw
materials is supplemented by manufactured products based on
low technology, and together enables them to earn sufficient currency to import more sophisticated products from technically advanced countries. The trend, then, is for both developed and underdeveloped countries to develop together and so enable money
to flow back and forth in mutual trade.

The need is for industry to stimulate consumption by offering an ever-widening choice through the provision of an increasing diversity of high-quality products. The growth of both developed and underdeveloped countries will have contrasting features which, again, encourage the total amount of activity. The
former may redeploy many of their unemployed in the services
sector without diminishing the absolute magnitude of the gross
domestic product. In the United States over 66 per cent, and in
Japan(15) 75 per cent, of the working population are employed

in various service activities. However, this trend may become less pronounced because the service industries will, themselves, also be subject to rapidly increasing productivity. For example, the introduction of new information technology into retailing will raise efficiency to unprecedented levels. Tesco, a leading retailer, is reported(18) to have committed some £100M of new sophisticated electronic point of sale equipment to its stores.

Invisible exports, frequently said to be a substitute for the wealth now created by manufacturing industries, is of growing importance. However, in the United Kingdom they are unlikely to pay for the import of manufactured goods. In 1983 a government Minister(19) stated that 'to replace the contribution made by the manufacturing industries to the balance of payments we should need to export £70 billion a year, which would, in turn, require our share of the world export of services to increase from under 10 per cent to over 50 per cent'.

The developing nations may be expected to increase both the absolute amount and the percentage of manufactured products in gross national product. Despite the promise of long-term advances in international trade, there will always be both winners and losers in a community and among countries. Fortunately, as the wealth of the world builds up there will be fewer losers, and their deprivation will be less.

There are many indications that governments, despite formidable obstacles, will adopt means of promoting general growth. There is now a wide consensus on the desirability of eliminating tariff barriers, quotas, and adverse exchange rates, while actions should be formulated to facilitate geographical mobility of peoples and products. Steps are being taken to achieve low or zero inflation since there is a wider recognition that, with weak governments, inflation leads to higher unemployment which initiates a cycle ending with hyperinflation. Zero inflation is associated with stable prices, smoothly running labour and financial markets, and the general adoption of meaningful investment criteria. Investment will be stimulated by low interest rates; stable prices will also remove one main source of acrimony caused by the difficulties involved in collective bargaining in the industrialized countries, and greater exchange rate stability is seen to be necessary for smoothly running international trade. Apart from a small minority, people have a deep longing for economic stability and are aware of its importance to governments and businesses. There is a long journey ahead. Although the European economic summit conference of 1975 at Rambouillet failed to fulfil even the most modest hopes, the 1983 summit at Williamsburg reflected a much greater accord among the participants. The seven heads of state agreed to work towards a greater convergence of economic policies, aim for a lower and more stable exchange rate, campaign against inflation, and reduce structural budget deficits. The five major economies agreed to monitor and, hopefully, to accelerate progress by means of a regular IMF surveillance committee. The current wish is for cautious expansionary fiscal or monetary policies which will not threaten an increase in inflation. Recovery must be based upon increasing output and higher employment rather than on higher wages and prices. If this and other desirable goals are not agreed, nations will have to resort to short-term palliatives. Inflation will be regarded as a device for discharging short-term debts, and massive government spending will be used as a means of counteracting a depression without regard to ultimate consequences.

Once confidence in the future is regained, and anxiety lessens, the hope is that polarizing attitudes will become less universal. The need is for a balance between rival claims for a free market, and a planned economy, so that the constraints and freedoms placed upon a community may be reconciled with individual aspirations. The behaviour of societies and economies can only be explained by numerous interrelated diagnoses, and the key to their understanding is the greatest of all challenges.

1.4 ADVANTAGES THAT ACCOMPANY GROWTH

Two observations made during company visits demonstrated that few smaller companies, still under control of the original owner or owners, wished to become medium-size, and few medium-size companies possessed either the means or the capability of becoming large. In an important contribution to the 1978 Wilson Committee, Mitchell(20) confirmed this general picture and reported that, following interviews with 50 medium-sized companies, it was apparent that the majority were more interested in maintaining their current level of activity than in expansion. Birley(21) has suggested that one common reason for firms wishing to remain small is that the ownership and management resides in the same person, or persons; so future company goals are determined not only by commercial considerations but by personal lifestyles and family considerations. Independence is often the entrepreneur's primary aim; it is not easily relinquished, and consequently a policy of minimum growth consistent with survival is chosen.

When discussing future plans with owners of smaller businesses, many additional reasons were given for rejecting growth. They explained that their work gave them a sense of fulfilment, they enjoyed sufficient material awards, and feared that expansion would attract attention from larger would-be competitors, and possibly unions, and, in any case, would be too risky in the current uncertainty of the United Kingdom economy.

Despite the apparent rationality of these views there are dangers from remaining small, and advantages from growing large. The dangers were vividly illustrated during the studies by several companies failing, or being acquired by larger firms. Small did not prove beautiful, but rather vulnerable, weak, and, in the end, insignificant. Many owners admitted that their reluctance to expand was based upon a hesitancy to introduce knowledge and skills with which they were unfamiliar, and so they would not appoint staff without whom expansion could not be safely undertaken.

Both small and medium size firms too easily acquired a false sense of security through a failure to carry out a systematic examination of likely risks. Perhaps, less surprisingly, they saw neither a social nor moral obligation to grow.

Manufacturers of scientific instruments, in particular, seemed content to supply one basic product to a single market sector, and were seemingly unaware of dangers from potential competitors, or from their customers failing because they, too, were dependent upon activities which were insufficiently diverse to withstand the recession.

Many reasons can be advanced in favour of small companies expanding. As owners recognize that shrinking markets eliminate customers and intensify competition, they should opt for growth and adopt policies which increase their chance of survival. They should examine competitors' activities, look for new market trends, scrutinize sales distribution patterns, and seek new outlets.

With growth a company may become an important entity within the local community, with its attendant advantages and disadvantages. As corporate knowledge of all matters concerned with products grows, its advice may well be sought by government departments on legislation concerned with safety, pollution, overseas trade, and other relevant topics. It will also, more easily, obtain representation on appropriate professional bodies. These responsibilities will carry both obligations and advantages and will, in turn, aid the process of growth. An established company is in a better position than one recently launched, to establish close and friendly association with both customers and suppliers. The large company may offer technical and market information in return for better products and service, and the new enterprise will wish to establish all possible links in order to remain in a growing market. An international reputation is a highly prized company possession, especially in the technical field, since continued growth is stimulated by interested companies making enquiries concerning licensing, acquisitions, and mergers.

1.5 GROWTH AND THE EMPLOYEE

Many who join a small company do so because they believe that they can identify readily with its aims. They are attracted by the possibility of informality and cooperation normally associated with a loose structure, welcome what they hope will be a short interval between the birth of an idea and its development into a marketable product, and are stimulated by the prospect of having to manage numerous and contrasting aspects of a business.

The reality may be quite different, and, in the course of time, most employees learn to recognize advantages which accompany expansion. It may free them from routine and mundane tasks to which they feel ill-suited, provide opportunities for specialised training, be seen as a means of providing new job opportunities, and constitute a step towards promotion with its greater responsibilities and wider horizons.

The achievement of growth is likely to have been founded upon those products that are capable of achieving worldwide renown despite vigorous competition, and will so be accompanied by a strong sense of self-fulfilment.

To remain competitive, staff in all functions will need to gain knowledge from outside their company, and they will be able to do this in a manner generally denied employees of small enterprises. There will be opportunities for international travel, for attendance at meetings of professional bodies, and international conferences. In time a company may well build up its own library and information service. The achievement of a successful reputation will bring its own rewards; it will not be difficult to discuss problems with one's peers and as experience grows so will the likelihood of sharing insight into problems with the more perceptive among those with whom collaboration takes place. A reputation for excellence also aids recruitment of staff of the highest ability.

Once a business has become established, and has achieved a measure of stability, employees will generally welcome expansion once they are certain that it can be advantageous to themselves, and not be accompanied by too many unknown consequences of change.

1.6 OVERCOMING OBSTACLES TO GROWTH

Extensive theoretical and practical studies of the innovation process, which have extended over some 15 years, have neither stimulated industrial growth nor reduced unemployment, and it is not surprising that the encouragement of new enterprises has proved difficult. Nor has a prevailing negative attitude been helped by studies which regard company size as a natural consequence of external conditions. Many economists appear content to observe that relatively few firms have sufficient resources for development (see, for example, Mitchell(20)), that only a minority of small firms can grow into larger ones, and there is a paucity of suggestions aimed at improving the initial processes of development. The first obstacle is thus an attitudinal one and will be surmounted when there is a widespread conviction that growth can be achieved by dynamism and ability in defiance of adverse external difficulties.

A resolve to expand is a crucial first step, but subsequent actions can be guided by an awareness of likely difficulties. Chapter 2 describes obstacles to both early and late stages of growth. However, a brief preview of the nature of problems and solutions may be helpful. An exposition of difficulties leads to the question of whether companies should always aim for indefinite expansion, and if they should, will future technologies and social changes help or hinder?

Concern is frequently expressed that the increasing industrial application of automation, and of robotics in particular, will be an obstacle to growth. The fear is that it will decrease the numbers employed and so reduce purchasing power. Whether or not this will happen remains a subject for debate.

Some 200 years ago there were dire forecasts that the introduction of looms would increase unemployment in the textile sector, while mass production of the Ford Model T and, more recently, the introduction of the main frame computer, were believed to augur similar misfortunes. In the event, reduced labour per unit of output produced lower prices, higher salaries, increased spending power, and resulted in higher not lower employment. Recent studies continue to give contradictory views (22, 23) and this is not surprising because many developments on which future growth is likely to depend are, as yet, unimagined. A likely outcome is that job losses will continue only while the industrial and commercial scenes experience rapid and great change. Thereafter, the world should face a more stable period in which considerable wealth could be produced(24).

That factors limiting the growth process will vary from company to company can be shown by considering two contrasting types of operation. The first example shows a company engaged on mass production of articles, free from technical complexity, in which there is a large production and a small administration staff; the second, a firm selling high added-value, science- based, products in a sophisticated market which employs a strong team of scientists and technicians supported by a small finance and sales section.

Growth of the first company will call for a large increase in production activities, but other areas, such as finance and sales, are unlikely to require a significant increase in resources. The first two obstacles, neither of which will be insurmountable, will be funding and the securing of a larger market. The third danger appears when expansion reaches a level at which a bureaucracy is created with its attendant risks of overmanning and low productivity. The smaller science based company with its specialized market may not be able to secure a large market, and emphasis will be placed on radically new products. Here, the chief obstacle will be centred upon the task of appointing scientists and technicians of sufficiently high calibre. Should this prove difficult consideration could be given to obtaining ideas or products from external sources (see Chapter 5). A further difficulty may be the sustaining of the original entrepreneurship. One solution which has been successfully practised is not to allow the small company unrestricted growth on the initial site but to expand into small autonomous units. These are placed under the control of an employee who has demonstrated entrepreneurial qualities, in the expectation that he will alone face the difficulties of initial growth and so gain the practical experience that is crucial for future company operations(25).

The slow rate of company growth in the United Kingdom has been attributed to industrial concentration into large units. In 1977 Jewkes(26) claimed that large firms have been wrongly condemned for administering rigid prices, creating price instability, retarding progress by suppressing inventions, and for evading social responsibilities. He further remarked that large size does not necessarily create a monopoly, nor is competition assured if firms are small. Penrose(27) has observed that there is no optimum, or even most profitable size of firm, and indeed, many of the most widely known companies, such as IBM and Matsushita, show no restraint in their expansion. The limitation is on the rate of growth. The progress of firms through their early growth cycles is described in Chapter 2, where it is seen that considerable time may elapse before owners or directors acquire sufficient experience, know how, and resources. Inadequate time is even a severe impediment in companies with seemingly large resources, and Beckett(28) has observed that even in the large-scale operation of car production, the setting up of an effective marketing operation takes three times as long as the time needed to develop a product and to build the plant.

In the context of growth and national prosperity, too little notice was taken of Jewkes' concern that of 60 000 manufacturing companies no less than 58 000 employed less than 500. Small companies employing less than 50 constituted a high proportion of the 7 000 bankruptcies and 13 342 company liquidations which occurred in 1983(29). Nevertheless, their collective contribution to the country's wealth is considerable. Established companies employing between 50 and a 100 could play an even more important role in wealth creation once they achieve their growth potential. They have two advantages: an experience of surmounting crises that accompany early stages of growth, and reasonable resources.

Future industrial growth may be stimulated by the new organizational patterns which are emerging as a result of new technologies, for example flexible manufacturing systems and robotics, and new attitudes to elaborate hierarchies and bureaucracies. One trend, already visible, is to split large disparate businesses into smaller, more manageable units. Macrae(30) believes that big companies will eventually fragment into mini-firms that will give more freedom to employees and so encourage entrepreneurism. He suggests, for example, that a company may form a typing services company from the typing pool and give the staff a fixed term index-linked contract with sufficient spare time for operating their own consultancy contracts. Small(31) comments that over the next ten years big factories will no longer be built, but will be superseded by a large number of small units.

F International is a noteworthy example of an imaginative approach to new ways of constructing a business(32) Founded in 1962 it is the largest computer software business in the United Kingdom. Its estimated 1983–84 turnover was in the region of £5m, and over the last four years the company has achieved a steady growth rate of 26 per cent. The organization has five regional United Kingdom offices, an engineering and science division, a subsidiary set up to serve microcomputer users, and three associated companies which serve the Scandinavian and Benelux countries, and the United States.

The workforce consists of over 760 highly skilled technicians and more than 150 administrators. The majority are freelance and work from home, but a salary career and promotion up to director level is possible. Contracts are fulfilled by teams which, ranging from two to 30 in number, are overseen by project managers. Entry qualifications include at least four years' experience in the data processing industry, not more than two years out of a job, and a willingness to work between 20 and 25 productive hours a week, with two days available for client visits. Over 95 per cent of the workforce are women with a family, although men are equally welcomed. The company offers freedom, a high degree of autonomy, and provides interesting and creative job and career opportunities. The success of the company suggests that it has met, and is meeting, a need caused by changes in both social patterns and the economic climate.

The above bird's eye view of the industrial scene leads to the paradox that a survival policy based on remaining small leads to failure, while a growth policy, at its worst, will result in survival.

1.7 SUMMARY

The continued growth within manufacturing industries is both a moral and socially responsible objective, for with it countries can achieve stability, strength and prosperity.

Britain's industrial revolution provided an early example of the wealth creation and distribution process, but unfortunately her early achievements were not sustained. A brief review of the successes and subsequent failures shows that a culture developed which was antipathetic towards industry. That similar difficulties were not shared by other industrialized European countries is largely attributable to the educational system in Britain at that time. The present need is for educational institutions to adopt an integrated role in the forging of a relationship between science, technology and society, and to create a climate in which both labour and management can work together towards a common objective.

The argument is advanced that in the medium to long term manufacturing industries and the associated service sectors will provide jobs for all who can, and are willing to work.

There is already an ever-increasing recognition by governments of the major obstacles to growth, and a wide consensus of the necessity to eliminate trade barriers, quotas, and inflation. International cooperation will be further strengthened by technical advances in communication and travel.

Underdeveloped and developed countries will interact and stimulate each other's capacity for expansion, and the sciences will make a major contribution. In developed countries the percentage of working population employed in service activities will exceed two-thirds of the working population without diminishing the absolute magnitude of the gross domestic product.

Not all small businesses wish to expand. In some instances owners are not motivated by commercial considerations, and in others a false sense of security is acquired. There is too little awareness that growth is an answer to the ever-present danger of shrinking markets in which customers are eliminated and competition intensified. People may join a small firm because they believe it will be possible to identify themselves with its goals, and they are additionally attracted by the prospects of becoming involved in all company activities. Others choose to work in large organizations in order to be free from the more mundane tasks and to have opportunities for quicker promotion to more senior positions.

A review of the means for obtaining industrial expansion has disclosed many diagnoses, but few precepts whose adoption would facilitate close control of the growth process. It is suggested that once the initial attitudinal difficulty is surmounted, attention be directed to the nature of the manufactured product, since obstacles to growth differ between products which are, or are not, characterized by technical complexity.

There is no optimum or even any profitable size of a firm. There may be limitations on the growth rate, but these may become less important with the introduction of new company structures that are being developed with the aim of being better able to adapt to changing social patterns and economic climates.

1.8 REFERENCES

1. Coleman, J. P. (1970). Creating economic wealth from technology, *Electronics and Power*, **August**, 285–90.
2. Adams, K. 'Engineering into the '80s', The *Production Engineer*, **July/August**, 33–5.
3. Smith, R. V. (1955). 'The task of modern industry', Federation of British Industries.
4. Carey, P. (1979). 'UK industry in the 1980s', *Proceedings, Institution of Mechanical Engineers*, **193**, No.41, 439–446.
5. Corlett, N, (1983). 'Why British industry needs more production engineers', *The Production Engineer*, **62**, No.9, 32–5.
6. Anon (1983). *EMF Fifth Annual Report on International Industrial Competitiveness*, European Management Forum, 53 Chemin des Hauts-Crêts, CH.1223, Cologny–Geneva, Switzerland.
7. Trevelyan, G. H. (1964). *Illustrated English Social History*. Penguin Books Limited, Harmondsworth.

8. Barnett, C. (1976). 'British history and its effect on attitudes in history', *IMechE Conference on Training and Career Development for Engineers.*
9. Parker, R. C. (1973). 'Science, technology and the changing role of the professional engineer', *Chartered Mechanical Engineer*, **20**, No. 3, 76–9.
10. Kaletsky, A. (1983). 'I do not see why short term rates should rise', *Financial Times*, **30 March**, 16.
11. Gabor, D. (1972). 'The mature society', *Nature*, **238**, 111.
12. Piatier, A. (1981). 'Innovation, information, and long-term growth', *Futures*, **October**, 371–82.
13. Dörfel, H. (1979). 'Innovation in the chemical industry', *12th International TNO Conference*, Rotterdam, 1–28.
14. Williams, E. (1982). 'The invention that shrank as it grew', *Financial Times*, **23 December**.
15. Bull, G. (1983). 'The re-making of Britain', *The Director*, **September**, 22–7.
16. Prest, M. (1983). 'Dynamic island states would benefit most from world economic revival', *The Times*, **28 February**, 18.
17. Taylor, B. (1980). 'Competitive strategies for world markets', *Chelwood Review*, **July**, 19–26.
18. Churchill, D. (1983). 'Technology is key to the future of retailing', *Financial Times*, **12 November**, 8.
19. Anon. (1983). *Mechanical Engineering News*, **January**, No.121, 1.
20. Mitchell, J. E. (1980). 'Small firms: a critique', *The Three Banks Review*, **June**, No.126, 50–61.
21. Birley, S. (1980). 'The large firm, small firm interface', *London Business School Journal*, **Winter**, 23–4.
22. Kaplinsky, R. (1984). *Automation - the technology and society*, Longmans, Harlow.
23. Anon. (1978) 'Social and employment implications of micro- electronics', *Central Policy Review Staff*, **November**, Cabinet Office, London.
24. Sinclair, C. (1983–84). Forum: Sir Clive Sinclair on Britain in the 1990s, *Journal of General Management*, **Winter**, 91–7.
25. Parker, R. C. (1983). Conference Report, 'The survival of industrial research and development', *R & D Management*, **13**, No.4, 261–4.
26. Jewkes, J. (1977). *Delusion of Dominance*, Hobart Paper, The Institution of Economic Affairs.
27. Penrose, E. T. (1959). *The Theory of the Growth of the Firm*, Blackwell, Oxford.
28. Beckett, T. N. (1977). 'Problems of size and efficiency in industry today', *Royal Society of Arts Journal*, **125**, No.5254, 624–36.
29. Fleet, K. (1984), 'Leaner–and now bigger', *The Times*, 17 May, 12.
30. Macrae, N. (1980). 'Into the age of the mini-firm?', *Chelwood Review*, No.8, **July**, 27–32.

31. Small, B. (1983). 'Factories of the future–what will they look like?',
    The Production Engineer, **62**, No.11, 30–1.
32. Symons, R. (1982). 'Today's technology tomorrow's world', *The
    Training Officer*, **September**, 248.

2

Growth and the Company Climate

2.1 THE EVOLUTION OF A BUSINESS ORGANIZATION

A business is launched when sufficient capital is raised, and the entrepreneur's ability to do this will depend partly on the nature of the invention, product, or service and the perceived balance between risk and the financial return, and partly on his ability to persuade people of his choice to join him. The likelihood of success, particularly in the initial stages, will depend upon the individual's acceptance of responsibility and recognition of opportunity, and the degree to which the group, acting as a team, seeks to improve its performance.

The growth process can proceed along many possible paths, and will reflect the manner in which the entrepreneur and his colleagues respond to the ever-changing nature of their climate. The growth of a business is a fascinating subject for study since it is conditioned by the whole spectrum of human behaviour. There are many parallels between growth of people and organizations, and for both the secret of achievement lies in the capacity to meet challenge with a response that retains the vitality of youth with the wisdom born of experience.

The way in which a company develops will depend, to a marked degree, upon personal relationships existing between those in control and their staff. It is therefore proposed to review that which is known about the factors which cause groups to behave in the way they do, and attention will be directed particularly to the early, and therefore crucial, stages of growth.

Few difficulties of a personal nature will arise when a small number of people agree to work together, for we are psychologically inclined to depend upon each other, and most people enjoy

21

working in groups towards clear goals. When a group's early efforts achieve success, both its sense of cohesion and desire to cooperate will grow. The value of teamwork is widely demonstrated by the numerous voluntary societies which do so much to sustain the fabric of our society. Not all groups, however, work well together: some achieve little, while others experience conflict among their members and eventually break up. It is therefore important to consider principles which should be adopted in order to increase the chance of success.

Group behaviour stems from the interaction of its members and it is necessary to begin by looking at the individual. Many sociologists use Bales'(1) set of categories for observing behaviour; this lists a number of personal attributes that can contribute to the success of a meeting. They include the demonstration of solidarity, the raising of others' status by offers of help and reward, humorous interjections and other means for releasing tension, a willingness to agree, understanding and compliance. Negative actions that hinder are: passive rejection, formality, tension, withdrawal, antagonism and self-assertion.

Mainsbridge(2), discussing the behavioural science aspects of groups, suggests that it is necessary to prevent dominance by any individual through reasons of expertise, personal attractiveness, verbal skill, self-confidence or the demonstration of access to restricted information, while psychological barriers to success have been described by Leavitt(3) as a reluctance to contribute through shyness, discouragement for fear of ridicule, and a feeling of inferior status.

Leavitt also states that a group whose members are in harmony with each other shows most freedom of thought, and little is gained in the long term by allowing a non-conforming individual to try to break a consensus of thought. Mainsbridge believes a consensual meeting is particularly good at innovation and at solving complex problems which arise when entering into unmapped areas of intellectual territory. He further states that a chairman should foster enthusiasm and high morale, and help to motivate, and try to conduct the meeting in such a way that members do not feel anxious about their progress. It is important to note that the sociologists, behavioural scientists, and psychologists all agree that each individual in the group must feel safe from conflict and aggression.

Group communication can be hindered by irrelevant information, termed 'noise' by the psychologist, and the reiteration of information, termed 'redundancy'. Some noise may be essential for creative thought in that it is an expression of personal need. An awareness of how to achieve success improves group communication and interpersonal relationships, but members must develop sensitivity towards each other's thoughts and feelings.

Because individual behaviour is frequently inconsistent and at times capricious, group behaviour is complex. Mills(4) has described seven classifications of groups, of which his cybernetics growth concept is the most suitable model for dealing with problems that arise in an evolving business (see also Normann(5), and Beer(6)). Here a group must seek to be continually aware of its performance with the aim of achieving self-improvement. The group tries to modify its environment in such a manner that its goal can be more easily reached. It must learn to adapt to change, to continually observe what is happening, and to seek an understanding of the actions of both the group and its individual members.

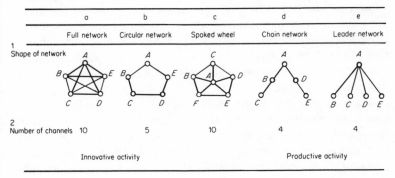

Figure 2.1 Types of communication networks, number of channels, and spheres of activity

Psychologists and behavioural scientists agree that the size of a successful group lies between eight and 15 people. This range permits the necessary diversity of ideas, allows members to contribute without long periods of unavoidable activity, and is not hindered by too much noise. Experiments have been carried out on the many different ways a group can be structured, and Leavitt(3) recommends a spoked-wheel structure

(Fig 2.1., from Jermakowicz(7)). It was found to be the least erratic, and can be effective in generating ideas provided that at the hub is someone who is skilled at guiding rather than at leading. It is recommended for small business groups during their formative stage.

A new business, even though it may only employ a few people, may continue for a considerable time without adding to its numbers. Eventually, new opportunities will arise for additional business and this will require new expertise or additional knowledge, and growth will begin. The original cohesive group will tend to break up and re-form into two or more formal, or informal, associations of employees. At this point the organization may be said to come into being, for there will be a need to integrate the work of the separate units in order to achieve the stated purpose of the company. The relationships between individuals and groups will comprise many strands, and a culture will develop and be unique to the company. There will be a realization that sanctions are beginning to appear, and some form of general discipline will be desirable. The culture will determine standards, values, patterns of behaviour, and may be responsible for initiating the exercise of authority. Under favourable conditions and good leadership it may transmute personal feelings into subjective values such as loyalty, duty, and sense of purpose; and keep within tolerable limits the negative feelings of envy, jealousy, and unbridled ambition.

An understanding of group development is helped by the concept of 'norms' which control the interactions between and within groups. A norm is, in effect, a group's consensus of how its members should act in various circumstances. In the case of a small business it will reflect the complex interactions between the initial vision and style of the founder, and the experiences subsequently gained by the group(8). Norms are rarely stated in an explicit manner and only the most perceptive members of a group will know that they exist and are influencing behaviour. An example of a norm that unknowingly modifies conduct is demonstrated when children invent a game and eventually imply what constitutes fair play. Taboos of primitive tribes are an example of norms the origins and purpose of which are often unknown. An example of an explicitly stated norm is when an agreed standard of behaviour is formulated through a byelaw.

It is useful to recapitulate those aspects of group behaviour which are relevant to a newly established business engaged in converting ideas into marketable products. The first expectation is that the team will work well together and lie within the optimum size range. Staff, at this early stage of growth, are unlikely to have acquired sufficient experience to deal easily with the inevitable succession of new problems, and so a high degree of adaption to new situations will be necessary. This can be best stimulated by observing the tenets of behaviour, outlined above, adopting a suitable structure, and arranging for feedbacks aimed at improving performance.

The first important growth phase will be signalled when numbers increase to a level that necessitates the integration of disparate groups. Unfortunately, this action is rarely taken soon enough because most executives tend to respond to events decisively, rapidly, and pragmatically. They are stressed by immediate pressures and so miss opportunities for standing aside in order to analyse attitudes and behaviour. It should be emphasized that unless there is a wider adoption of a more analytical approach to factors which are responsible for the performance of staff, companies will fail to reach their optimum level of competitiveness. To show that precepts of group behaviour can be of practical help, their application to a clearly identified subgroup within a large organization will be described. As befits a book on growth through product innovation, the illustration will concern the direction of a research division, although similar reasoning can be applied to other functions.

A starting point is Maslow's work(9) in which he describes a number of needs which, if not satisfied, lead to stress. The two basic needs which are probably most applicable to research staff are esteem and self-actualization, and from these can be derived personal needs of achievement, mastery, competence, status, recognition, self-fulfilment, and the urge to do creative work to the limit of one's capacity. As discussed by Parker(10), a research director should consider whether these important personal needs are likely to be inhibited by 'norms' associated with the business(11).

The important scientific norms are a belief that:
(1.) judgement should be based on traditional, impersonal, objective criteria (universalism); (2.) there should be no

proprietary rights on knowledge gained from basic research (communality); (3.) advancement of knowledge is more important than personal gain (disinterestedness); (4.) the work should be subject to independent validation (organizational scepticism) by one's peers; and (5.) a researcher should select his problems (autonomy).

The traditional norms in industry are quite different and involve the acceptance of: (1.) company loyalty; (2.) comformity to established policies and procedures; (3.) some degree of hierarchical authority; (4.) secrecy of company expertise; and (5.) monetary reward based on status.

The likely areas for mismatch between norms are many, and unless the research director resolves them and so provides an appropriate environment for the function he controls, he will not easily attract and retain men of eminence. For example, the company norm of secrecy is in opposition to the scientist who wishes to gain recognition from his peers by publishing papers, have freedom to attend and contribute to lectures and conferences, be given opportunities to gain postgraduate degrees, be encouraged to serve on committees and professional bodies, and to participate in joint programmes of research with universities and research associations. The director must convince his colleagues of the error of lessening a scientist's output by giving him administration duties, and argue the case for reasonable freedom of communication.

It also follows from the category of needs that a scientist's status within the organization manifestly derives from his scientific ability and not from seniority. Duties must not be tied to rank, and personal growth implies acceptance of a flexible attitude towards tasks. It is not by accident that there is a trend for research and development divisions to be split up into separate groups comprising between eight to 15 members. The structure comprises a research director having under his control a number of creative groups each led by a manager, and served by service groups headed by second line management. (See Chapter 8, p.191). In all businesses there is this continuous need to observe and analyse activities in order to initiate action aimed at avoiding situations which might lower effectiveness. Chief executives in small businesses, and top management in large companies, should try to achieve an understanding of employees' aspirations, reactions and behaviour. This will be made easier if future situations can be linked to specific growth patterns.

Mention has been made of an early discontinuity in the first stages of company development when leadership is required to integrate the activities of the emerging disparate group. Many other examples of abrupt changes associated with growth have been identified.

A noteworthy contribution by Greiner(12) is shown in Fig 2.2. There are five specific phases of growth: creativity, direction, delegation, coordination, and collaboration. Each stage, other than the first, is both an effect of the previous phase and a cause for the next, and all are heralded by signs of an impending crisis. Greiner points out that the implication of his analysis is that each phase of management action is narrowly prescribed if growth is to occur.

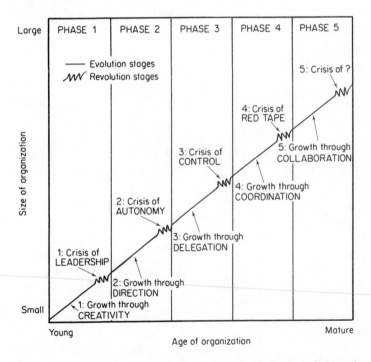

Figure 2.2 Growing organizations—the five phases of growth.

Table 2.1   Entrepreneurial characteristics

	Small business	Medium-sized enterprise	Medium-large enterprise	Large-scale enterprise
Objective	market determined intuitive improvising creative	market directed *ad hoc* strategy intuitive creative	market directed broadly strategic inventive quickly adaptable	market dominating systematical strategical managerial mechanism according to model
Management	personal authoritarian direct	personal consultation with those concerned incidental use of specialists outside the enterprise	personal small specialist team use of specialists outside the enterprise	team large specialist team line organization and staff organization function structure consultation structure
Product	strong personal market feeling quick reaction great flexibility no planning	strong personal market feeling quick reaction great flexibility short-term planning	moderate personal market feeling delayed reaction reasonable flexibility short-term and irregular long-term planning	no personal market feeling very slow reaction no flexibility long-term planning

	no market research	product and market information by literature and courses no market research	incidental market research	market research
Staff	personal relation boss—employee one large family large involvement	personal relation employer–employee labour community some involvement	hardly any personal relation cooperative relationship some involvement	no personal relation—only in groups— cooperative relations no involvement (only at the top?)
	motivation easier little trade union influence no staff council	motivation easier little trade union influence no staff council	motivation more difficult moderate trade union influence staff council	motivation very difficult large trade union influence staff council
Location	personal choice	personal choice	combination of personal choice and business economic and social factors	team choice after elaborate examination
	relocation fairly easy to realize	relocation realizable as a rule	relocation difficult to realize	relocation hardly to be realized
Real property, if any	saleable as a rule	moderately saleable	difficult to sell	hardly saleable
Financing	family bank	family bank	family bank open market	bank open market

Velu(13) simplifies the hypothesis by reducing the five phases to three: namely, the pioneer, the differential, and the integrated stages.

In the pioneer stage, the instigator of a business has a clear perception of what he wishes to achieve and therefore tends to be autocratic. The small, close-knit, staff will be fully informed upon progress, and communication will not be a problem. Activities will be directed mainly towards increasing profitable sales, and, because the number of customers is likely to be small, staff will be able to improvise and so meet customer requests for special attention. As business progresses, staff will acquire additional experience in bookkeeping, and will exercise greater control over cash-flow. They will probably face growing competition and become increasingly aware of a need to strengthen their hold on the market. A stage will be reached at which a crisis may be threatened unless further staff are appointed to strengthen existing knowledge, expertise, and skills. This action will signal a need to set up a more formal organization.

The differentiated stage begins with the introduction of a scientific professional approach to business management. Bookkeeping will now be but one activity within an administration that will involve the control of finance, stock, salaries and wages. Planning, budgetary and evaluation systems will appear, and as growth rate increases, the organization will expand to comprise many separate departments and divisions. As time passes a number of the new groupings will seek power and, no doubt, gain considerable autonomy which may threaten friendly cooperation among staff. It is important to be watchful for these change signals of the next transitional phase.

For companies whose growth is sufficient to enter the integrated stage, their main concern will be to maintain a growth rate through the introduction of a succession of new products. They must at all costs avoid the stagnation which too often accompanies maturity. A principal danger will be the emergence of a rigid bureaucratic organization which will inhibit teamwork across the many disciplinary and functional groups. As discussed in Chapter 8, (p.195) the virtues of the smaller organic structure must be preserved, otherwise motivation and organization will be destroyed and decline will be inevitable. As Velu(13) remarks, 'management, in the integrated stage, should move from the top of the organization structure to the centre, when the board and senior managers will be increasingly concerned to improve information, communication, and integration'.

The above two classifications cover growth from the beginning to maturity. For a small business growing to medium size, Velu proposes a complementary approach in which the following six thresholds are listed:

1. Starting
2. Liquidity
3. Delegation
4. Management
5. Success
6. Succession.

and when proceeding to a higher threshold the entrepreneur is advised to consider whether it would be advantageous to call for professional advice. Reference is made to seven crossroads in the development of an enterprise at which financial and legal advice is necessary. The seven phases are:

1. Formation of an enterprise
2. Extension or relocation of the enterprise
3. Import and export business
4. Change in the legal status
5. Cooperation and merger with other enterprises
6. Takeover of an enterprize
7. Succession in or liquidation of an enterprise.

Table 2.1 is a useful summary by Velu of the way certain company characteristics tend to vary with the size of an enterprise. A second task for top management is now clear. It is to develop an awareness of growth patterns, and have the prescience to recognize impending signs of an approaching crisis.

2.2 OVERCOMING BARRIERS TO INNOVATION

Once a firm foundation for a company's future is established, it becomes increasingly important to undertake product innovation.

Numerous obstacles have to be identified most of which recur, albeit in a different guise, from year to year.

Goodwin(14) defines three input barriers which have to be surmounted if an innovation is to succeed. Problems must be recognized, reported, and dealt with. The probability of overcoming all three barriers is, of course, the product of the individual probabilities, and if these are as high as 0.8, 0.95, and 0.7, respectively, the overall probability is only 53 per cent.

In addition, Goodwin lists a further seven internal obstacles. Three of these—time, energy, and an ability to show interest in an innovative idea—he describes as barriers to action. The remaining four, termed organizational barriers to output, are board approval, money, resources other than money, and customer acceptance.

In Fig. 2.3 are displayed the probabilities of surmounting all ten barriers when the probability for each is 0.95, 0.7, and 0.5, respectively. The overall chance of success is seen to drop from three in five to one in 1000. The inference is that competent and creative staff are needed to control the work involved in all ten barriers. Excellence, however, is sparsely distributed and it seems beyond the reach of most companies to achieve a probability for all ten barriers as high as 0.75, and as this only yields a general probability of approximately one in 20 it is surprising that the chance of success is rarely quoted as worse than one in five. There are two possible explanations. The first is that teams, rather than individuals, are assigned to each of the more difficult barriers and these, in turn, arouse the interest of their colleagues so that probabilities increase in proportion to the total effort.

The second is that when a problem is seen to be intractable, a concerted effort is directed along several parallel paths. The degree to which cooperation takes place is a vital element of success, and will depend upon the company ethos and the form of the organization.

Witte, in a series of important publications(15) discussed the concept of barriers as early as 1972. He saw the development of new products as a process which passes from an initial stable situation to a final stable state via a series of intermediate unstable ones. It is the first move from the existing order that causes the greatest disturbance. Staff become anxious because they realize

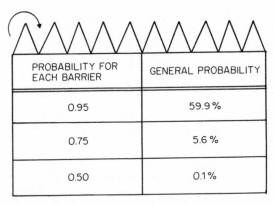

PROBABILITY FOR EACH BARRIER	GENERAL PROBABILITY
0.95	59.9%
0.75	5.6%
0.50	0.1%

Figure 2.3 Surmounting barriers to innovation

that resources are finite and hence existing plans will have to yield to new ones. Their action will, though perhaps unconsciously, result in pressure being generated to preserve the status quo. Once this initial apprehension of change is overcome there remains the danger that organizational and behavioural patterns which have been abandoned will re-form in a new guise, and seek to establish new conditions. It is for this reason that development momentum must be sustained. Witte sees the first goal of a business organization wishing to innovate as the overcoming of inertia caused by fear of change, and he describes it as the barrier of will. It is akin to Goodwin's barrier to action.

The second obstacle is the barrier of capability, and arises when finance, research, marketing, manufacturing, and many other functions (reference Goodwin's organizational barriers) need to adapt to new circumstances. Many activities will interact and give rise to complex technical, organizational, and social problems. Considerable difficulties will be experienced when product development calls for staff with new disciplines which may be beyond the ability of existing staff. Witte postulates that both barriers can be best overcome through the personal commitment of staff members who are able to supply the necessary energy and act as 'promoters'. Though differing in detail this role is essentially that of a 'product champion'. The concept of a series of obstacles is illustrated in Fig. 2.4.

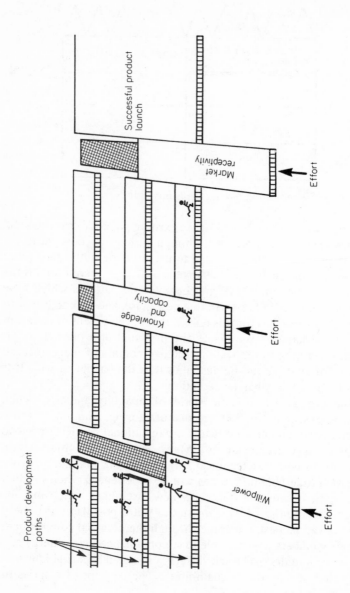

Figure 2.4 Obstacles to successful product launch

The barriers of will are overcome by a 'promoter of power' who uses his authority within an organization to advance an innovative idea. He may hold legitimate power to protect those in favour of innovation and exercise sanctions against opponents. The latter should rarely be necessary, however, if conditions favourable for growth, namely, an acceptance of change, are encouraged by the board. The most important single challenge of our time is how to cope with the accelerating changes now occurring, and Nolan(16) has explored ways in which it is possible to manage change to the advantage of everyone concerned.

The barriers of capability are destroyed by a 'promoter of know how' through his specialized knowledge to see that all new and complex tasks are quickly mastered. Unlike the promoter of power he will not rely on power and authority and, indeed, is likely to be a member of line management who is strongly motivated by an innovative idea. He will rely on knowledge to influence colleagues, and will continue to study and instruct others. Most members of an enterprise could act as a promoter for certain innovations, provided they are capable of infusing colleagues with enthusiasm. A scientist who conceives an idea may not necessarily possess this necessary quality.

Witte's studies have shown that because the two promoters pursue their goals by different means success rarely attends the merging of both functions in one man. Success more often follows when a power promoter and a know-how promoter work together in close cooperation. The two promoters can be of different rank but while cooperating for the purpose of innovation they act as equals.

2.3 THE EXERCISE OF POWER

It has been argued that (1.) good relationships between departments and divisions can be fostered by those in authority acquiring a sensitive perception of staff behaviour; (2.) knowledge of typical growth patterns can help staff to anticipate future problems; and (3.) the many known barriers to innovation can be surmounted by possession of sufficient know-how and the exercise of power.

Good personal relations and an anticipation of discontinuities in the growth process are particularly necessary in the early stages of a company. With further growth the emphasis passes to decision making, and the allocation of responsibilities for action by those who possess the necessary power to wield authority.

There are many views on the nature of power, but the one which seems the most in accord with the author's experience of industry is the strategic-contingency theory of Salancik and Pfeffer(17) which sees power as something accruing to organizational subunits (individuals, departments) enabling them to cope with critical organizational problems. The salient point is that power and the exercise of authority, based on status, will not be exercised satisfactorily unless there is tacit agreement by interested staff that actions taken will be to their ultimate benefit.

The principle that power in a free society has to be earned, rather than conferred, explains why attempts by cliques to form coalitions to institutionalize power, and so diminish the authority of others, are rarely long-lived, unless the environment is so stable that new critical contingencies do not arise. An hypothesis of the coexistence of power with an ability to cope with critical problems is demonstrated in the evolutionary pattern of professional institutions (Parker(18)). They derive power to exercise their authority from so ordering their affairs that members alone have the skill and knowledge demanded by their clients. In contradiction to the above general observation, the professions have shown that their power can be fairly successfully institutionalized and this has been done by the virtual elimination of competition. They have also sought stability by drawing up codes of practice the aim of which has been to ensure that knowledge is available to all their members through their learned societies and meetings, to prevent the standing of members being jeopardized by colleagues, and to disown proprietary attitudes. Although many of the professions were founded over 300 years ago their structures have survived extremely well. There are emerging signs, however, that their monopoly is beginning to be challenged as a result of recent economic and social change.

2.4 RESPONSIBILITIES OF EXECUTIVES FOR GROWTH

The need for a board to create an ethos favourable to growth has been highlighted by Witte's observation that a power promoter has authority to exercise sanctions against opponents of innovation. Should such action be contemplated it is a sign that conflict has reached a level at which it is channelling energies away from constructive work. It is therefore proposed to examine the implications of Sections 2.1, 2.2, and 2.3, on the specific responsibilities of management.

The first duty of a board is to define the purpose of a business and to list the important goals. The second is to formulate appropriate strategies and translate them into operational plans. While these operations are the responsibility of the company executives it is always wise to invite managerial staff to participate and to explain the reasons for important decisions. If the majority of employees feel, instinctively, that executive plans will create a company in which they wish to work, then the board's power will be reinforced and so be better able to cope with changes that accompany growth.

The greater the power and authority of executives the easier it will be to so delegate functions that the managers will be seen as suitable trustees for authority. As an hierarchical power structure evolves with changes in the company climate, care should be taken to see that it is done in a manner which appears logical and therefore acceptable.

Executives must adopt attitudes and actions which will reassure staff that they recognize that company growth, whether sought from new products or through the acquisition of know-how, will be associated with risk. Cannon(19) has aptly and convincingly pointed out that occasional failures are inevitable because if the chance of success with a new venture is 1 in n, then $n-1$ failures may be expected before success is guaranteed. A company's belief in growth can be additionally demonstrated by designing reward systems that look to future potential rather than past achievements.

The one characteristic which distinguishes a growing from a static business is the high degree of staff understanding and the involvement needed to deal with change. Staff consequently expect competent direction, and the first responsibility of a chairman must be to appoint colleagues of the highest calibre. The intellectual abilities of executive directors should, ideally, be equal; otherwise there could be the temptation for a board member to gain an advantage for his division through an ability to present arguments based on a superior intellect.

Directors should in turn only appoint managers of high ability since a few people of excellence can easily outweigh those of many times their number who possess only average ability. A few people of excellence will usually determine the success of a business. Satisfactory answers to problems may call for a threshold of creative ability, while the best solutions are usually found to call for the least possible resources. There is a need to establish a culture in which elitism is recognized as an essential virtue.

Few boards would dispute that they have a responsibility to be seen as a compatible group, and those who are able to work with a high degree of unity must have a significant advantage over the majority that experience intermittent conflict.

The obstacles to good relationships among members of a board are many and are to be expected, since the spectrum of their responsibilities causes a high incidence of overlap. A common cause of disputes is the competing claims for resources among divisions, and the correct balance between long-term and short-term needs. For example, a works manager may wish to spend a considerable sum upon means for reducing manufacturing cycles, and finds it easy to gain the support of the directors responsible for sales and finance. Coincidentally the research director may decide that the only way of maintaining the company's international technical lead is to introduce new scientific disciplines by the appointment of additional staff. His case may be supported by the export director whose experience has been that it is easier to obtain overseas licensing arrangements when the company has an international reputation for technology. Even though a board may adjudicate between the opposing claims for funds, unless it is seen to be done in a convincing manner, arguments will continue at lower levels of management. If conflicts of this nature become frequent power becomes fragmented and morale quickly falls. It is a matter of regret that so little regard is paid to current studies aimed at reducing antagonism by making groups more open to change.

While conducting studies on the management of innovation within companies, it quickly became apparent that harmonious relationships between directors prevailed when the chief executive displayed high qualities of leadership. These companies were the most successful. Outstanding leaders are readily recognized and appreciated, and, in the companies headed by them, their virtues were usually commented upon by all levels of management, although not always directly.

Characteristics of leadership are a sense of mission, a vision of the future, wise judgement, and a conviction of how to attain clearly expressed goals(20). The above discussion on power also suggests that leadership devolves from a sensitivity to those environmental changes which are likely to alter critical contingencies, and an ability to explain the need for change to employees in a convincing manner. To be able to discharge these obligations without bias a chief executive should not align himself to a particular function and, if his earlier experience was centred on one function, a comprehensive training programme should have been undertaken. The qualities of leadership are probably innate, but it is reasonable to assume that they can be developed through training and example. An important indication of leadership will be the ability to resolve conflict through conviction rather than authority.

The necessary traits of a good leader differ among contrasting personalities and cannot be easily described. Prominent among the qualities, however, are integrity, commitment, ability demonstrated through achievement, capacity for work, and approachability.

An indication of weak leadership is the relinquishing of power to ambitious staff who are always alert to opportunities for gaining authority. The transfer of authority through such means could be exercised for good or ill, but should be avoided for the reason that the process is fortuitous. A chief executive should guard against the possibility of one individual gaining a disproportional degree of power since it will discourage those from whom the power has been taken, and, possibly, give rise to a coalition of enemies who will then overreact and cause confusion. Too much, and, too little power vested in the heads of divisions and departments can greatly reduce the effectiveness of a business organization.

2.5 ESTABLISHING A CREATIVE ETHOS

Even if staff have good personal relationships, and are well directed, changes that inevitably accompany growth will raise important problems and upon the quality of their solutions will depend the company's future. Toynbee(21) expresses this with his customary clarity, and wrote, 'To give a fair chance to potential creativity is a matter of life and death for any society. This is all-important, because the outstanding creative ability of a fairly small percentage of the population is mankind's ultimate capital asset.'

Witte draws attention to the importance of problem solving when he states that the promotor of know-how has to direct specialized knowledge in order to master quickly new and complex tasks. From this it must not be inferred that achieving a solution is a mechanistic operation that can be rapidly executed, dependent only upon the availability of specialized knowledge. The reality is that optimum solutions are most consistently produced by creative people working in a sympathetic environment. A few comments on the nature of creativity are given in Appendix I, (p.42).

Consultancy sessions aimed at stimulating new ideas for business opportunities are discussed in Chapter 7 (p.144) and give a practical illustration of the importance of stimulating creativity when seeking new business opportunities.

2.6 SUMMARY

The growth of a new business reflects the manner in which the entrepreneur and his colleagues respond to challenge. In the early stages they will enjoy good cooperation through a mutual awareness of a common goal, but their chance of success can be heightened by referring to studies on the nature of interpersonal relationships. Fewest difficulties will be experienced when numbers are between eight and 15, and a structure should be adopted which will enable this group to monitor and so improve its performance.

The need to appoint staff with special experience signals the beginning of the first growth phase, and a team which was once cohesive is likely to split into two or more associations. It is at this point that an organization comes into being. An increasing degree of formality will be accompanied by the emergence of constraints, and a business will develop its own culture. The need for leadership will become apparent and it will become advisable to study the patterns of employees' aspirations, reactions, and general behaviour.

Many stages of abrupt change associated with growth have been identified, and an analysis of them helps both the anticipation and avoidance of likely crises.

An examination of difficulties encountered when innovating suggests that the probability of success is very low unless cooperation is effective and the company assigns more than one team to the more difficult problems.

Attention is drawn to the barrier of will needed to overcome the reluctance of staff to accept change, and the barrier of capability needed if all functions are to adapt satisfactorily to new circumstances. Both can be surmounted by making two appointments: the first to be invested with the necessary power to advance innovative ideas, and the second with sufficient specialized knowledge to see that new and complex tasks are quickly mastered.

Power is seen to accrue to an individual, or group, by virtue of a proven ability to deal with critical organizational problems. The ability will be based on knowledge, and may be augmented by gaining unique practical experience of certain aspects of a company's activity. Power and the exercise of authority is more readily acceptable when staff realize that it is, and will be, used to their ultimate advantage.

To create an ethos favourable to growth, a board should state the purpose of the company and define the goals. Senior staff should be encouraged to participate in the formulation of strategies and their translation into operational plans, in order to be seen by all employees as suitable trustees for authority. Executives and senior staff must be seen to welcome innovation, with its attendant risks, and will ensure that positions of crucial importance are staffed with people of the highest calibre.

Success is attendant upon harmonious relations between executives, and this calls for a chief executive with high qualities of leadership. A company's future is seen to depend upon a few outstanding creative minds, and a brief review is appended on the nature of the creative process.

APPENDIX I
THE NATURE OF CREATIVITY

Creativity is said to be involved when attempts that are made to satisfy a need can be obtained neither by behaviour based on reflexes or habit, nor by discriminate choice within the area of things learned, such as simple algebra. In the present context creativity, whether conceived with solving problems or searching for new opportunities, is an activity which is relevant to new situations requiring original and imaginative concepts.

The first phase of the creative process is the recognition of a problem (whose identification may be difficult) and the reception of a stimulus; and the second is the formation of a concept(22). Haefele(23) suggests that a solution requires the marrying of two or more apparently unconnected concepts aided by insight. The formation of the required concepts may demand training in one of the appropriate disciplines and indeed depends upon the availability of knowledge, the use of analogies, and other types of information. Insight may involve memory recall through evocative stimuli, and frequently requires a long period of gestation termed incubation. If the problem is difficult, incubation will need to be preceded by thorough preparation and seems to require periods of intense thought alternated with periods of relaxation.

There are numerous recorded examples where solutions to problems have suddenly come to individuals who have not been consciously thinking about them. Hadamard's book on 'The psychology of invention in the mathematical field'(24) is a rich source of cases, few of which are more explicit than Gauss(25), when referring to an arithmetical theorum which he had unsuccessfully tried to prove for years, writes 'Finally, two days ago, I succeeded, not on account of my painful efforts, but by the grace of God. Like a sudden flash of lightning, the riddle happened to be solved. I myself cannot say what was the conducting thread which connected what I previously knew with what made my success possible.'

Some writers have defined as many as seven crucial steps for problem-solving, but it is generally agreed that the four basic steps are preparation, incubation, illumination and verification. Psychologists have found that the third of the above steps is not a straightforward process and Kleinmuntz(26), in particular, has demonstrated one important block to creative thought. It occurs when a well practised method for solving a particular class of problems is applied to a problem for which it is not suitable. It is termed the '*Einstellung*' effect and strongly inhibits the generation of new concepts largely because the process is a subconscious one. Stenhouse(27) has, more recently, suggested that the ability to postpone an immediate instinctive response to stimuli is crucial to creative thinking, since only then is the mind free to scan its past experience in order to generalize and synthesize.

If a large number of concepts of differing kinds are necessary to solve problems it suggests that flair allied to a vivid imagination are preferable to a high degree of logic and a dull imagination. This explains why so much emphasis nowadays is placed on divergent thinking which is believed to correlate with peoples' fluency in creating hypotheses. Here one is reminded of Brunelle's(28) statement that education does not necessarily help creativity since it may put too much emphasis on order, clarity, and logic, and ignores the value of irrational thought, imagination, and all that tends to be wild and chaotic in nature. An apt synthesis of these views has been given by Brunelle who states that problem solving is a conscious or subconscious manipulation of ambiguous symbols in a systematic pattern so as to produce new meaning.

The number of concepts would be expected to increase with the diversity of knowledge, perception and relationships between ideas, and this is why a group is likely to perform better than an individual.

A large element of the above hypothesis is part of everyday experience. Crossword addicts, among countless others, will accept that incubation takes place, for how otherwise can solutions become apparent during unexpected moments? A business executive may not welcome the thought that creative ideas cannot always be produced on demand, or that irrationality can, at times, be more important than logic.

As will be seen in Chapter 7, methods have been developed to raise the level of creativity, but they are unlikely to be adopted unless positive action is taken to stimulate a creative ethos throughout the company. This can be done by executives encouraging staff to use every available means for obtaining ideas, not excluding the non-traditional ones.

2.7 REFERENCES

1. Bales, R. F. (1950). *American Sociological Review*, **15**, 257
2. Mainsbridge, J. J. (1973). *Journal of Applied Behavioural Science*, **9**.
3. Leavitt, H. J. (1951). *Journal of Abnormal Social Psychology*, **46**, 38–50.
4. Mills, T. M. (1967). *The Sociology of Small Groups*, Prentice Hall.
5. Normann, R. (1977). *Management for Growth*, Chichester, John Wiley & Sons.
6. Beer, S. (1981). *Brain of the Firm*, Chichester, John Wiley & Sons.
7. Jermakowicz, W. (1978). 'Organizational structures in the R & D sphere'. *R & D Management*, **8**, Special Issue, 107–13.
8. Schein, E. H. (1983). *The Role of the Founder in Creating Organizational Culture*, Organizational Dynamics Seminar, 13–28.
9. Maslow, A. (1973). *The Farther Reaches of Human Nature*. Harmondsworth, Pelican Books.
10. Parker, R. C. (1976). *The Nurturing of a Creative Atmosphere in an R & D Laboratory*. Institution of Mechanical Engineers, London, 1–22.
11. Blume, S. S. (1974). 'Behavioural aspects of research management—A review', 40–76.
12. Greiner, L. E. (1972). 'Evolution and Revolution as Organizations Grow'. *Harvard Business Review*. **July/August**, 37–46.
13. Velu, H. A. F. (1980). *The Development Process of the Personally Managed Enterprise*. European Foundation for Management Development's 10th European Seminar on Small Businesses, 1–21.
14. Goodwin, H. B. (1977). 'Overcoming organizational barriers to innovation'. *Symposium - Research on Research*.
15. Witte, E. (1973). 'Organisation für Innovationsentscheidungen— Das Promotorenmodell Decisions'. Schwartz. Göttingen.
16. Nolan, V. (1981). *Open to Change*. MCB Publications Limited, Bradford.
17. Salancik, G. R. and Pfeffer, J. (1977). 'Who gets power—and how they hold on to it: A Strategic–Contingency Model of Power'. *Organizational Dynamics*, **Winter**. AMACOM (American Management Association).
18. Parker, R. C. (1973). 'Science, technology and the changing role of the professional engineer', *The Chartered Mechanical Engineer*, **20**, No. 3, 76–9.
19. Cannon, C. G. (1981). 'Industrial research and profitability', *Physics Bulletin*, **32**, 320–2.
20. Parker, H. (1979). 'Advice to a new chairman'. *McKinsey Quarterly*, **Winter**, 2–20.
21. Toynbee, A. (1967). *On the Role of Creativity in History*. University of Utah Press.
22. Gagne, R. M. (1959). *Annual Review of Psychology*, **10**, 147–172.

23. Haefele, J. W. (1962). *Creativity and Innovation.* Reinhold Publishing Corporation.
24. Hadamard, J. (1954). *The Psychology of Invention in the Mathematical Field.* London, Constable.
25. Gauss, K. E. (1886). Letter mentioned in *Revue des questions scientifiques*, 575.
26. Kleinmuntz, B. (1966). *Problem Solving: Research, Method and Theory.* Chichester, John Wiley & Sons.
27. Stenhouse, D. (1974). *The Evolution of Intelligence.* London, Allen & Unwin.
28. Brunelle, E. A. (1971). *Journal of Creative Behaviour,* **5**, No.37.

3

Growth and the Environment

3.1 EVOLUTION AND INNOVATION

In the last chapter attention was drawn to changes which businesses experience during their early stages of growth. Based upon an understanding of the manner in which staff respond to change, it was suggested that crises could be avoided by taking anticipatory actions in order to maintain steady progress. This chapter is concerned with the effect of environmental change on technical development and demonstrates that competitiveness can only be sustained if an understanding is gained of those factors which influence the future. Growth can be achieved in many ways, but they all involve either or both of the two contrasting philosophies of evolutionary development and innovation. It is therefore appropriate to make a clear distinction between the two approaches. The two terms will be defined, illustrated, and some indication given of their likely interaction with the environment.

Evolutionary development describes work aimed at continuous improvement to meet changing market needs, and may be applied to both processes and products. It is likely to involve experience strengthened by a continuous assimilation of evolving science and technology and is normally aimed at sustaining or expanding the share of existing markets.

Innovation involves the creation of a new idea, often an invention, together with its progression to the marketing of a new material, process or system. It implies a discontinuity sufficiently great to merit an examination of its possible effects on the company's strategies, structure, and attitudes.

The first of the two definitions is being increasingly replaced by the terms 'incremental' or 'continuous innovation'. Both imply progression by small steps, and this is misleading. Innovation contrasts with evolution in that it is a radical operation, is rarely free from risk, and demands exceptional skill and determination if the degree of change is to be successfully dealt with.

Examples of both innovation and evolution can be found in the friction materials industry which is approximately 100 years old, and has its roots in a typical entrepreneurial invention. In an examination of the manner in which it sought answers to problems over many decades Parker(1) observed that because its more traditional products comprised a large number of constituents, it became increasingly difficult to isolate the effects of changing any one; and until the late 1960s, technical developments relied upon empirically based methodologies. Progress became asymtopic and each new advance required more and more effort at an ever-increasing cost.

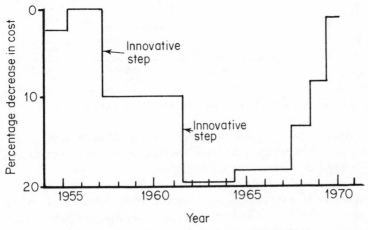

Figure 3.1 Effect of innovation on the cost of a clutch facing

Although this pattern generally applied, there were notable exceptions. They were highlighted by plotting the cost against time of several popular qualities of friction materials that were manufactured in large quantities. Fig. 3.1 shows how the cost of an automobile clutch facing changed over the period 1954–71 and Fig. 3.2 shows the increase in the cost of a brake lining over the same period. The important feature is the two sharp discontinuities seen in the first figure. Reference to earlier records

showed that the two instances of a dramatic fall in the cost of
the clutch lining were attributable to serendipity. Further sharp
discontinuities were sought, and found, in records of other products. Research staff were intrigued when shown the contrasting
nature of innovation and evolutionary development, and major
changes were made in the direction and structure of the laboratory.

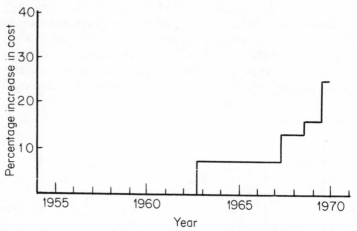

Figure 3.2 Variation in the cost of a brake lining since 1955

Nyström(2) has published a classification of both an evolutionary (which he terms 'positional') and an innovative company, and
his findings are summarized in Table 3.1.

The characteristics suggest that a successful positional company rarely experiences an unstable environment and, consequently, eschews change and adopts a reactive attitude towards
market requirements. The manufacturing operations associated
with relatively unchanging products are concerned with attain-
ing the highest possible efficiency and this they achieve by giving
attention to automation, group technology, rationalization, and
other production engineering procedures. If research and development, or design, is carried out it is normally evolutionary and
defensive in character. The company organization, not unnaturally, tends to be impersonal, hierarchical, and status dependent.
The boards of positional companies can afford to lay stress on financial control and corporate plans may, with some justification,

be based upon the use of extrapolated historic trends of financial statistics.

Table 3.1   Some characteristics of positional and innovative companies

Function	Positional company	Innovative company
Board	Emphasis on financial control	Innovation-orientated— future perceived as uncertain
Organization	Impersonal, hierarchical, status-dependent	Dual structure— vertical and horizontal
Marketing	Reactive—stability based on attractiveness of product. Closed marketing strategy	Contructively creates an unstable environment
R & D	Defensive, evolutionary	Aggressive, innovative
Production	Efficiency, rationalization, and long runs	Openness to change

The innovative company accepts that the future is likely to be uncertain and the process of corporate planning must itself generate new insights, be dynamic and reiterative. The marketing function may even seek to expand its sales by causing, and feeding upon, a degree of chaos. Manufacturing operations will thus need to be designed with a view to the possibility of introducing new production methods. These will be more important than efficient production. Research, design and development will be important activities and will be aggressive and innovative rather than reactive. The organization pattern is likely to have the character of a dual structure: a vertical based on a specialized function, and a horizontal concerned with coordinating independent activities.

Because more and more attention is being focused upon innovative enterprises which exploit new technologies in the expectation that they will provide employment for those made redundant by basic industries in decline, it is salutary to note that there are still industries in which evolutionary development is the better choice. An example is one in which both customer and supplier are apprehensive of disasters which could be caused by radical change. Hidden dangers that can unsuspectingly accompany intended improvements are real, and an illustration can, again, be taken from the friction materials industry.

In a paper on the development of railway brake blocks Parker(3) has described the issue as one of maintaining safe train operation in the face of uncertainties produced by strong interactions between a railway brake block, wheel, the condition of the rail, and rain. Repeated applications of a brake block against a railway wheel may so alter the composition and roughness of the wheel and rail surface that the train's braking behaviour is adversely affected and, to an extent, that depends upon the percentage of coaches and wagons the brakes of which have been fitted with the new brake block material. The effect is insidious, difficult to detect initially, and depends upon the weather, so that neither test machine assessment nor service trials can predict future behaviour with an acceptable degree of certainty. Even when all preliminary tests and trials seem satisfactory past experience dictates that a new material is only slowly introduced, and it is not unusual for several years to elapse before even a minor change can be judged completely safe. Evolutionary change is therefore often preferred for systems when complex, long-term, interacting variables make assessment unreliable.

Evolutionary development may so strongly attract those who wish for a stable external environment, in order to avoid undue stress, that they may subconsciously discount the associated danger of zero growth or even company liquidation. The challenge of innovation will be for those who know that the existing order may have to be destroyed in order to build anew.

In practice, most companies have to deal, simultaneously, with three classes of possible future products: the long established, those under development, and those at the development stage. A fortunate circumstance occurs when a significant proportion of a product range is capable of sustaining its market share over a reasonable period through evolution. It provides a stable platform from which innovations can be launched. If a mature industry can meet foreseen changes in the external environment in a planned and gradual way, it has the best opportunity of designing an organization, and creating an ethos, which can cope both with stability and change.

One of the most disturbing changes in the environment occurs when a company of apparent strength and stability is suddenly faced with competition based on a new technology. Cooper and Schendel(4) who studied the response to technical threats in 22 separate firms within seven competing technologies, namely:

steam locomotives *vs* diesel engines
vacuum tubes *vs* the transistors
fountain *vs* ballpoint pens
solid fuel power *vs* nuclear power plants
safety *vs* electric razors
aircraft propellers *vs* jet engines
leathers *vs* pvc and porometric plastics,

highlighted two important weaknesses: companies did not recognize the true significance of external threats soon enough, and did not maintain a strong competitive position.

The picture was complex, and though many of the results were tentative, there were three observations on possible pitfalls. The first concerned the tendency to underestimate a threat when first perceived because the technology was in its initial crude stage—a judgement which is more likely when a company can look back on earlier threats which it has survived. The second was the loss of a valuable lead, caused by a decision to defer action until sales of products, based on existing technology, declined. The third danger arose when a forecaster failed to understand differences in needs of market segments and so did not relate these to probable improvements in the new technology.

In the seven industries studied by Cooper and Schendel it was seen that improvements were sought and reached after the new competing technology was marketed. For instance, the smallest and most reliable vacuum tubes ever produced were developed after the introduction of the transistor. Furthermore, all but one of the 22 companies continued to make heavy commitments to improve the old technologies, and no threatened firm adopted a strategy of early withdrawal to concentrate on the new.

The above examples show that there is no one easy path to growth, because, in a changing environment, means must be found to perceive sufficient of future needs in order to learn how resources can best be allocated between evolution and innovation.

3.2 ENVIRONMENTAL FORECASTING

In Chapter 2 it was hypothesized that a business could expand in a smooth and steady manner, provided sufficient was known about both staff attitudes and the causes of those management crises which were characterized by Greiner (see p.27). It was also inferred above that the discontinuities shown in Figs. 3.1 and 3.2 need not have happened, had the research and development director recognized earlier the importance of innovation. The question to be answered is the extent to which abrupt changes in a business can be avoided.

Holroyd(5) has observed that even large scale disturbances may only seem to occur suddenly because our perception and understanding of inter-related phenomena are limited, and cites, as examples, the eruption of a volcano and a revolution in a far distant country.

We are apparently insensitive to coming change because it proceeds in stages that are so small that they tend to escape observation; however, once detected, slowness has the advantage that it provides time for measurement and analysis. While the ultimate consequence of change can be often foreseen, as witnessed in the successful prediction of earthquakes, there will be cases when this is not so. An assassination, caused by a stray bullet in a revolution may have worldwide consequences, but even if physical constraints governing its path were not too immediate to permit analysis, the presumption of free will would make the situation indeterminate. Catastrophes do occur, but the term should be restricted to those classes of events which are mathematically indeterminate.

A not unreasonable assumption would now seem to be that most events which affect business are a consequence of gradual change. This being so it should be increasingly possible, as experience and knowledge accumulates, to either predict or modify future events. While a detailed discussion of forecasting is beyond the scope of this book, a review of its more important aspects will be useful at this stage.

The general term, environmental forecasting, refers to the process of predictions, all trends and occurrences external to organizations which, together, play a major part in determining future events. Such predictions are frequently classified under one or more of five headings, namely: economic, technological, political, social, and ecological. Forecasting calls for an understanding of past, and a knowledge of current, events and involves their integration through judgement and intuition.

It is now widely acknowledged that weather forecasting is reasonably accurate, while during national parliamentary elections millions accept predictions based on a few early returns. Economic forecasts are also widely reported, but are subject to considerable scepticism.

Economic forecasting employs four basic tools of science: observation, postulation of hypotheses, prediction, and validation, and is concerned with probabilities of apparently random occurrences. The components of economic forecasting have been discussed by Ramsey(6), who concludes that a prediction usually comprises a mixture of naivety, intuition, and a scientific approach. The latter, econometrics, is a means of using mathematical models and statistical techniques to test theories, and the computer is enabling a very large number of complex, interacting variables to be processed.

Forecasting techniques are commonly classified under two headings: extrapolatory and normalized (see, for example, Twiss(7)). The former includes those techniques based upon an extension of the past, through the present, for the purpose of anticipating the situations to which they may lead. In so far as extrapolatory forecasting includes technical and economic components the methods are usually quantitative. The normalized approach is predominantly subjective and begins with future goals, needs, and desires, and the many aspects are traced backwards to determine how the endpoint may be achieved. A combination of both exploratory and normalized approaches is generally practised. A choice of suitable forecasting techniques has been confused by the listing of over 100 different versions in the literature (see Jantsch(8)). Fortunately the complexity has been lessened by a classification made by Jones(9) which provides a working framework. Methodologies are classified according to their contribution to the following four primary elements:

Qualitative: A narrative description of future events which, based chiefly on intuition and memory, nevertheless includes systematic procedures.

Quantitative: Technical measures of performance, efficiency, or some other measure of a product, together with economic indices as exampled in market share.

Time: Time in the future when a qualitative description and quantitative data become effective and real.

Probability: Chance, expressed in statistical terms or as a percentage, that those forecasts based on the above three elements will materialize.

One of four examples given by Jones(9) to illustrate the type of prediction obtained by applying these techniques is that the expectation of life for humans will be longer (qualitative), and achieve 90 years (quantitative) by 1995 (time) with a probability of 20 per cent or increased to 99 years by the year 2000, with a probability of 10 per cent.

Important constituents of the qualitative elements are social forces. Catling(10) clearly substantiated the contribution of qualitative factors in determining the course of product innovation in the textile industry. Instances of cause and effect have been observed between: (1.) general affluence, central heating, easy care, and lightweight clothes; (2.) educational trends and the diminution of high-craft industrial processing; and (3.) the rising expectations of labour and the gradual disappearance of jobs which entail physical discomfort and danger. Because of the large number of social factors which could influence a proposed innovation Catling suggests that attention be directed under five headings: education, industrial norms, social norms, status in society, and physical amenity.

The subject of environmental forecasting is too wide to be dealt with in this chapter, and comments will now be restricted to some of the technological forecasting techniques of practical relevance to growth based upon product development.

3.3 TECHNOLOGICAL FORECASTING

The aim of a business is to gain advantage over competition, and Catling and Rodgers(11) recommend technological forecasting as a means of introducing new processes and products which will take competitors unawares and make their plant and machinery obsolete overnight. By integrating knowledge over a wide range of subjects, forecasts should emerge that will enable a firm to manage future adverse events, ensure a new technology will work, and develop new products which will be accepted by customers.

In a 1972 conference report on 'Technological Forecasting and Long Range Planning', the wider view was expressed that technological forecasting will improve the interaction between technology and society, and increase cooperation between industry, government, and international organizations. The conference considered that a formalized application of the techniques was likely to be very costly and noted that industry was often sceptical of its value. The subsequent 12 years has, however, seen significant advances; and firms engaged on high technology, with long lead times, realize increasingly the need to foresee future markets.

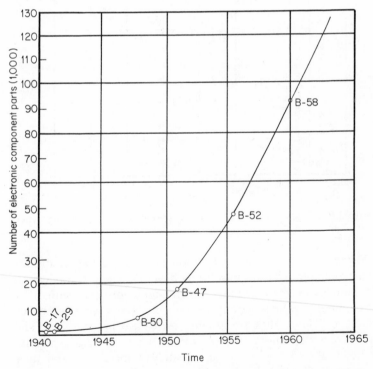

Figure 3.3 Trend of complexity of electronics in AF weapon systems

The rationales for technological forecasting have been described by Bright(12). The first is that data which characterizes product performance exhibits a systematic behaviour over time and through growth to yield a recognizable pattern, and that abrupt major deviations rarely occur. The second rationale is that technology responds to needs, opportunities, and the provision of resources. The third is that new technology can be forecast through studying the process of technical innovation: 'Coming technology casts its shadows far ahead, and signals of emerging technological innovation can be identified and monitored, ultimately providing the basis of a forecast'.

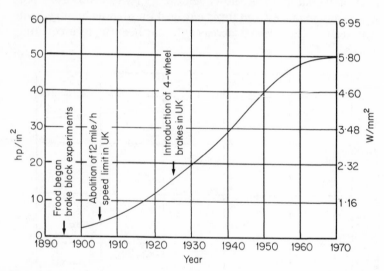

Figure 3.4 Increase in peak energy dissipation of a brake lining. since 1900

Wissema(13) looks upon technological forecasting studies as systematic investigations into future development and application of technologies, in order to see which interactions exist with other developments, which actions are possible, and what effects they will have. The studies are usually carried out pragmatically and have a relatively short time-horizon of five to ten years, and are limited in the scope of the object of study. Longer-term and more complex studies are usually referred to as futurology (see Kahn, Brown, and Martel(14)).

Among the numerous references to successful outcomes of technological forecasting, a classic case is described by Bright whereby one of the most important developments of the 20th century was indirectly accelerated. A study carried out by the Aironics Laboratory, US Air Force, in 1952 logged the number of electronic components in five successive military aeroplanes, and an extrapolation of the data to the B-50 forecast that it would contain 90 000 to 100 000 elements (see Fig. 3.3). From the known failure rate of vacuum tube circuitry then in use, it was statistically unlikely that all systems would remain operational long enough to support flight, and that an alternative solution to electronic components had to be found. This, and other analyses, led to the Air Force funding $10m for the development of solid state circuits and thus enabled Texas Instruments to provide the first integrated circuit.

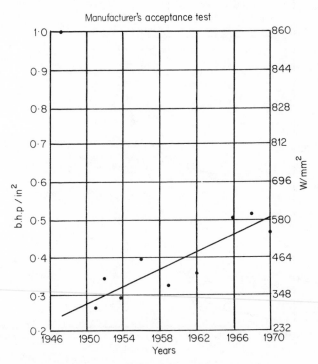

Figure 3.5 Increase in maximum engine b.h.p. per unit area of clutch facing. since 1947

Forecasting techniques based upon extrapolating trends in numerical data have, traditionally, been used as part of the general armoury in research, development and design laboratories. Fig. 3.4 (see Parker(1)) is an example taken from the friction material industry. It shows how the peak energy dissipation per unit area of brake linings has varied over the period 1900–70. The curve is typically sigmoidal and flattens out between 1964 and 1970.

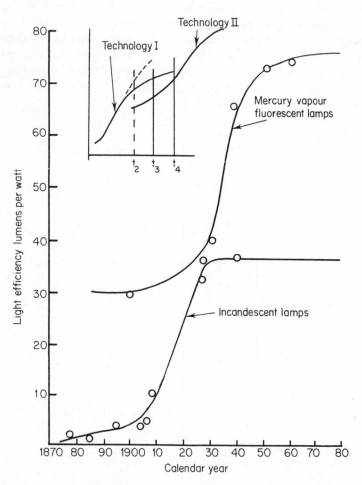

Figure 3.6 An example of two technological escalations: the light efficiency of incandescent lamps and mercury lamps against time

Figure 3.5 is a plot of the duty imposed on an automobile clutch facing over a 24-year period and indicates a linear increase of 100 per cent in the maximum brake horsepower per unit area. An interesting feature of this plot is that, while the unusual (and unintended) high overload acceptance test of 1946 may have incurred unnecessarily costly research and development, the resulting product continued without modification for over 20 years and maintained an 80 per cent share of the market.

Though many companies have experienced unsuccessful as well as successful uses of technological forecasting, it remains a most useful tool provided care is exercised in choosing the parameters and the time period does not extend beyond five to ten years into the future.

Figure 3.6, based on Wissema's paper(13), is an example of two technological escalations: the light efficiency of both incandescent and mercury lamps against time. The S-shape is attributed to slow progress in the early stages during which learning and experience overcome bottlenecks, followed by initial exponential growth as learning proceeds, and finally a slowing down as a ceiling is approached. The dotted line inset represents technological succession in which the two curves overlap, and is a potential cause of serious misjudgement because between t_2 and t_3 the potential of the new technology is underestimated.

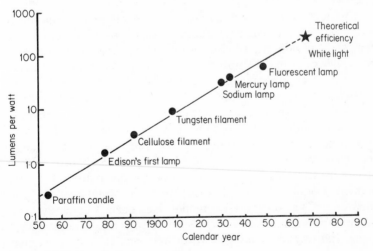

Figure 3.7 Technological succession

Fig. 3.7, again taken from the same paper, illustrates technological succession by plotting the point at which technologies reach their maximum performance against time. Unfortunately, instances of such orderly progress over so long a period of time are rare.

Fisher and Pry(15) have developed a model which represents a frequent occurrence, namely, the substitution of an old by a new technology in order to meet a given need. Their simple formula for substitution analysis is:

$$\frac{f}{1-f} = \alpha f (1-f) \tag{1}$$

where f is the degree of substitution, t is the time, and α is a constant indicating the slope of the substitution curve. It assumes that if the substitution has progressed by a few per cent it will proceed to completion, and that the fraction of the newly substituted is proportional to the remaining quantity of the old.

Integration of the above formula gives:

$$\frac{f}{1-f} = \exp \alpha (t - t_0) \tag{2}$$

which means that $f/1-f$ plotted against time on a semilog scale gives a straight line.

This technique often provides a reasonable picture on which to base a decision when substitution ranges from one to five per cent and provides a good forecast over the range 5–20 per cent. At higher rates of substitution subsequent progress is more easily found through other means. In Fig.3.8 is shown an example given by Bright of the substitution of steam for sail. The 1830 forecast was optimistic, the forecast made in 1860 was reasonably accurate.

When a competitor's product is to be directly attacked by a substitute, formula (1) above will be relevant, and sales plotted against time on a linear scale should be S-shaped. When stages in sales history show a reproduceable pattern they are termed product life cycles (PLC), and are often used for prediction. The substitution model has been frequently validated, although Cooper and Schendel(4), have noted exceptions. Existing technology may, for example, expand rather than

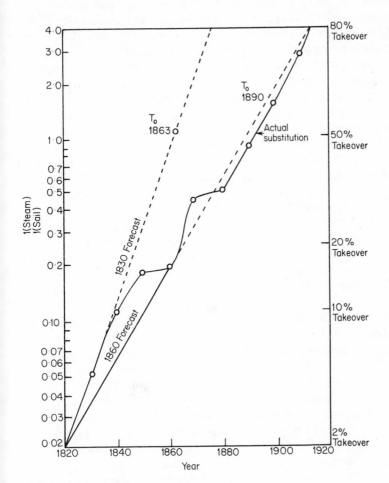

Figure 3.8 Substitution of steam for salt

decline when challenged by a new, growing technology (safety razors *vs* electric razors, p.51). Again, when a new technology caused a decline in traditional products it could open up new markets unavailable to the old (fountain pens *vs* ballpoint pens).

Numerous PLC patterns have been cited in the literature and range from the bell-shaped form, which corresponds to the 'adoption curve'—shown by products which are manufactured to satisfy a short-wave fashion and are not replenished—to more complex curves in which adoption cycles are followed by a sequence of repurchase cycles (see Midgley(16)).

Though PLCs have an appeal due to their simplicity there is considerable scepticism on their practical value. Day(17) discusses five basic issues, any one of which may introduce sufficient ambiguity to negate the concept. The two most important are the definition of the product market, and those factors which can influence different stages of the life cycle. By way of illustration, the second factor would be operative if there was sudden change caused by new legislation, fashion changes, and international trade barriers. The main doubt arises because product sales are rarely a function of time alone.

Abell(18) emphasizes the importance of precise definition and, as an example, defines a product as the application of a unique technique to the provision of a particular function for a specific customer group. A high technology product, and household consumer goods, match this definition and their sales behaviour frequently conforms to a PLC pattern. Many papers recommend that both the subdivision of a market and its customers will increase the chance of obtaining meaningful curves, and one company gained success by plotting growth in different geographical regions for customers when income lay within specified limits. Though success may follow careful application and development, the subject requires more examination.

In attempting to anticipate competitors' likely product innovations a board of directors may be meticulous in gathering and studying information, but in its interpretation intuition is likely to play a dominant role. This is as it should be since this particular insight or sixth sense, though largely subconscious, reflects numerous observations over a long period, based upon experience and knowledge. Although a board should, together, be more successful than one of its members acting alone, the advantages of a larger number of skills, wider knowledge and experience, may be partly vitiated by reason of conflicting interpersonal relationships (see p.38).

In the 1960s Helmer and Dalkey at the Rand Corporation developed the Delphi technique for the purpose of making more effective use of expert opinions. It involves circulating questionnaires to a list of suitable experts, who remain anonymous, and may number from tens to hundreds. The process is, essentially, reiterative and consists of three stages:

(1) A request is posted to the respondents for a prediction of the date of occurrence of certain specified events. On receiving the replies analysis is made of the data, showing the medium and interquartile range of replies.

(2) The statistical data is sent to the experts who are requested to reconsider the information and, if necessary, make a new prediction. If this falls outside the interquartile range a reason should be stated. The statistics are again returned to the sender who retabulates the information.

(3) On receiving the results of the second stage the experts should consider whether a new prediction is called for, and if it is, he should state counter arguments to the reasons advanced by other experts.

The Delphi method has been demonstrated to introduce into organizations future policies which had hitherto been ignored. Although this technique is commonly used in connection with product developments it can, of course, apply to fields other than technology.

Although the Delphi method is a powerful way of collecting opinions of experts it may be difficult to identify, contact, and secure their help. Considerable analysis and correspondence is also inevitably involved and the operation may constitute a major task. Success will depend upon the way in which the forms are structured, and the questions must be precise and allow neither equivocation nor misinterpretation. A number of researchers have described modifications to the technique, and four interesting models are briefly described by Bright(12).

Attempts have been made to use the Delphi technique to gain an indirect commercial advantage. A large number of respondents are selected who are experts and, additionally, are in a position of authority. The expectation is that, by reason of their position in industry, commerce, government or academia, their conviction of the rightness of a prediction will conspire to bring it into being. This is reminiscent of Holroyd's(5) 'creative future', which is the outcome of actions based upon dreams and explorations.

3.4 SHORT-WAVE TRADE CYCLES AND LONG-WAVE ECONOMIC ACTIVITY

When the annual value of the trade index is plotted against time over a period of years it will almost certainly show some signs of non-random fluctuations. If a pattern is obtained which approximates to a periodic wave form, it suggests that this, too, could serve to warn of future external environmental changes and so allow a business to anticipate and act upon favourable or

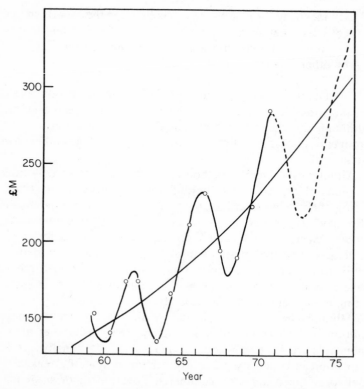

Figure 3.9 Fixed capital investment in the chemical and allied industries (at 1963 prices)

unfavourable occurrences. Whether or not this can be done depends upon how uniform is the observed periodicity.

In a paper on trade cycling in the United Kingdom Lunt[19] notes that many business statistics exhibit a combination of cyclical component and a longer term growth trend, and, in illustration, plots the annual expenditure from 1959 to 1970 on fixed capital investment in the chemical and allied industries (Fig. 3.9). In order to highlight the cyclical component he abstracted the continuing growth curve mathematically to give Fig. 3.10, in which the vertical axis represents the magnitude of the deviation of the annual value from the smooth curve trend. In graph C the amplitude of the cyclical swing exceeds the slope of the trend curve so that the effective growth rate ranges from +25 per cent to −15 per cent per annum. At the time his paper was written Lunt believed that consistency in trade cycling patterns over the 15-year period justified extending the curves three to four years ahead. He suggested that it could serve to indicate when resources should be husbanded in order to prevent reductions in labour and capital expenditure during the troughs.

The four to five year cycle had wide credence from 1958 until 1970, but Brittan[20], in a critical review of the literature on business cycles, pointed out that for most indicators a trough occurred in 1975, after an interval of only three-and-a-half years, and was then followed by a six-year cycle.

It is therefore not surprising that trade cycles also show regular patterns over short periods because both business organizations and the economy form adaptive systems termed 'closed loop', and may become stable when appropriate adjustments are made to the appropriate inputs. An example of actions taken to stabilize one aspect of the economy is the use of money controls to stabilize the bank rate.

The fluctuating nature of outputs, found in many engineering systems, has been subject to theoretical analysis under the discipline of cybernetics, and the consequent understanding has been a crucial factor in many spectacular achievements.

A cybernetic system has one very important property. If actions taken to change the behaviour of a system cannot be made immediately, the observed output will oscillate and its magnitude will depend both on the number of stages involved and their associated delays. Unfortunately the collection of statistics may take many months, decisions may require lengthy consultations, and implementation of agreed actions may also be necessarily slow. The cycle of an oscillation is a fundamental property and this will, in general, increase with the time needed to feed in new inputs and also with some system characteristics such as the degree of non-linearity.

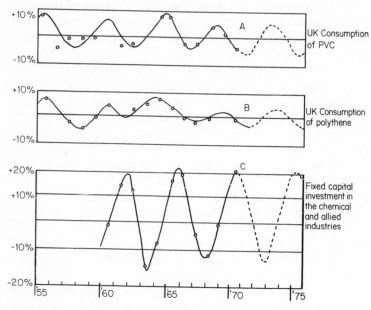

Figure 3.10 Typical cyclical fluctuations

Most businesses would operate more efficiently in the absence of oscillations, and the question is whether or not action can be taken to avoid them. In engineering systems oscillations can be virtually damped out by applying the required degree of anticipatory control at the critical moment, and of a magnitude that is proportional to the rate of change in the systems output. These conditions are so demanding that the continuous oscillations observed by Lunt are to be expected as a consequence of our ignorance of the almost limitless interactions of economic factors. The regularity of the observed cycles, however, suggests that the economic controls which were being exercised did not effect drastic change. An interesting non-mathematical description of cybernetic principles has been given by Brix(21).

Long-wave economic cycles have been observed, and are associated with Kondratieff(22) who plotted an index of consumer prices against time. Fig. 3.11, attributable to Smith(23), shows the location of the three-and-a-half cycles so far identified whose period varied between 40 and 50 years. This curve has, not unnaturally, attracted world-wide academic interest, and reference should be made to Ray(24) for a review of 42 references on the subject.

Numerous suggestions have been made to explain the long-term wave and, irrespective of their reality, a brief reference to the more significant hypotheses may be useful to those concerned with forecasting possible future events.

Simiand(25), in 1931, developed the concept that a period of rapid growth based upon an expanding technology alternated with a period when capital was spent on improving existing technology, and inefficient enterprises were being eliminated.

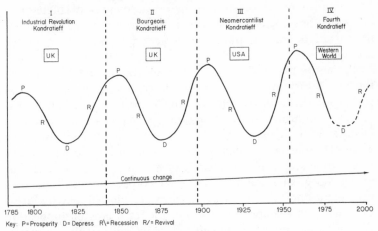

Figure 3.11 Kondratieff cycles after Kuznets, Schumpeter, Mensch

Schumpeter(26) later suggested that stagnation of an economy which had attained apparent maturity had to be destroyed by innovative creativity when further economic development opened up new avenues of growth. He attributed the long wave to the bunching of major innovations in production methods. Many writers have used hindsight in order to identify which innovation could have caused the Schumpeter waves, and a thorough analysis has been given by Weston(27). Her contribution is particularly valuable in that it directs attention to the varying timespans between an invention, its innovation, and its widespread adoption. Mensch(28) has been concerned to develop a theory which could explain how innovative activity is initiated in times of technical change with the object of influencing economic decisions. Forrester(29), who was involved in the development of large-scale computer simulation models of the US economy, observed a 50-year cycle between the growth and collapse of the capital sector while unaware of the literature on the long wave. He suggests that when an economy is on the upturn, capital investment grows to the point of saturation and exceeds future needs. The decline in investment which must follow, results in rising unemployment, rising prices, and high interest rates. During this stage innovations become unpopular, social structures are attacked, and the depression persists until creative management instigates the replacement of obsolete plant and failed businesses. Innovation is thus seen not as an initiation of growth but as an important reaction to an upturn in the economy.

The above hypotheses have common strands, and all underline the vital contribution of innovation to economic development. A search for short-term cycles by smaller firms may produce patterns of significance when attempting to collect facts and opinions in order to anticipate external environmental changes; but long wave phenomena are mainly the concern of large organizations whose time-scales involve studies of macro economics. There is, as yet, no great regularity in the 40–50-year periods, and they vary according to which relationships are plotted and the areas to which data are applied. The need is for a greater application of cybernetic knowledge to subjects which will enable economies to be controlled and, when appropriate, stabilized.

3.5 A PRACTICAL VIEW OF FORECASTING

Econometric and input–output models are among the most successful, but their high cost largely restricts their use to large groups. The need of smaller companies to obtain foreknowledge of changes to the environment is equally great, and their approach must necessarily be simpler. Once a product and market have been selected for examination a beginning can be made by studying the economics of the product, the process, and the relevant industrial sector. A scenario for two to five years ahead can be constructed and compared with the current situation. Since the techniques described in Section 3.3 are relatively straightforward, a search can be made for a relationship which, in the absence of a known ceiling, may be extrapolated and so open a small window on to the future. Figure 3.12 (see Chambers, Mullick and Smith(30)) is a reminder that there is an optimum forecasting cost and, hence, care should be taken to avoid unnecessary sophistication. These authors rate three forecasting techniques capable of achieving fair to good, or very good, accuracy over a period of two or more years ahead. Two are qualitative and one quantitative, namely, the Delphi technique, market research, and trend projection.

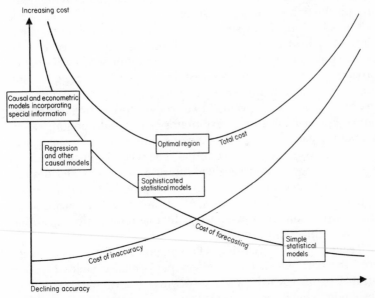

Figure 3.12 Cost of forecasting versus cost of inaccuracy for a medium-range forecast, given data availability.

Although forecasting is not yet firmly established nor widely accepted as a subject in its own right, many of its elements are incorporated in normal business practice. Increased use of computer data banks should increase the scope of possible methodologies and will facilitate the sub–division of market segments. The need is for a new stimulus, and this could be provided should an examination of company reports show that board decisions, aided by forecasting techniques, had an above-average success rate.

3.6 SUMMARY

Competition demands an understanding of the effect of environmental changes on technological development. Growth, based upon product development, can involve both evolution and innovation processes. The latter contrasts the former in that it is characterized by discontinuities which are sufficiently large to evoke resistance to change. Nyström(2) classifies companies as either evolutionary or innovatory and describes how the functions of their boards, and marketing, research and development, and production departments differ.

For a business to expand in a smooth and steady manner environmental disturbances must be forecast, and, because change occurs only slowly, it is necessary to develop sensitivity to this process. Environmental forecasting refers to these trends and occurrences external to an organization which play a major part in determining future events. Two types of forecasting techniques are recognized. The first—exploratory—is based upon an extension of the past in order to anticipate future situations, and is usually expressed in quantitative terms. The second—normalized—is mainly subjective, and specifies future goals, needs and desires, and traces aspects backwards in order to see how the endpoint may be achieved. The many methods are classified according to the contributions they make to qualitative information, quantitative data, time, and probability.

Technological forecasting is based upon three rationales which are: the data which characterizes performance yields a recognizable pattern which rarely shows major deviation; technology responds to need, opportunities and resources; and new technology can be forecast through studying the process of innovation. Forecasting techniques are based upon extrapolating trends in numerical data and can be of considerable value, provided the parameters are carefully chosen and the time period does not exceed five to ten years. Mention is made of the S-shaped curve which may result when the volume of sales is plotted against time for a product in which a substitution of a new for an old technology is taking place. Reference is also made to the Delphi method of forecasting which is concerned with harnessing collective opinions of experts.

Short cycles of less than ten years' duration are often observed, and although the extent of their stability is not yet known, they may repay study. A great deal of attention has been paid to trade cycles of approximately 50 years but there is, as yet, no real basis on which to formulate policy, since the period is irregular and not well understood.

More sophisticated techniques, as exampled in econometric modelling and scenario writing, are the most successful but are beyond the reach of all but the larger companies. The search for simpler relationships which can be extrapolated is worthwhile and often opens a small window onto the future.

3.7 REFERENCES

1. Parker, R. C. (1970–1). 'The art and science of selecting and solving research and development problems', *Proceedings, Institution of Mechanical Engineers*, **185**, No.64, 879–93.
2. Nyström, H. (1979). 'Creativity and Innovation', Chichester, John Wiley & Sons.
3. Parker, R. C. (1976). 'The importance of service trials in the development of railway brake blocks', *South African Mechanical Engineer*, **26**, No.2, 43–52.
4. Cooper, A. C. and Schendel, D. (1976). 'Strategic responses to technological threats', *Business Horizons*, **February**, 61–69.
5. Holroyd, P. (1979). 'Some recent methodologies in future studies: a personal view', *R & D Management*, **9**, No.3, 107–116.
6. Ramsey, J. B. (1977). 'Economic forecasting models or markets?' *Hobart Paper No.74*, The Institute of Economic Affairs.
7. Twiss, B. (1980). *Managing Technological Innovation*, Longman, London, p.215.
8. Jantsch, E. (1967). *Technological Forecasting in Perspective: a Framework for Technological Forecasting, its Techniques and Organisation.* OECD, Paris.
9. Jones, H. (1975). 'A systematic approach to technological forecasting'. *R & D Management*, **6**, No.1, 23–30.

10. Catling, H. (1972). 'Conditions for innovation - with particular reference to textiles', *R & D Management*, **2**, No.2, 75–82.
11. Catling, H. and Rodgers, P. (1971). 'Forecasting the textile scene: an aid to the planning of a research programme'. *R & D Management*, **1**No.3, 141–6.
12. Bright, J. R. (1977). *Technology Forecasting as an Influence on Technological Innovation: Past Examples and Future Expectations*, Symposium on Industrial Innovation, University of Strathclyde, Glasgow, September, 1–21.
13. Wissema, J. G. (1982). 'Trends in technology forecasting'. *R & D Management*, **12**, No.1, 27–36.
14. Kahn, H., Brown, W. and Martel, L. (1976). *The Next 200 Years*. Associated Business Programmes (Hudson Institute), London.
15. Fisher, J. C., Pry, R. H. (1971). 'A simple substitute model of technical change'. *Technological Forecasting and Social Change*, **3**, 75–88.
16. Midgley, D. F. (1976). *Innovation and New Product Marketing*. Croom Helm Ltd., London.
17. Day, G. S. (1981). 'The product life cycle: analysis and applications issues'. *Journal of Marketing*, **Fall**, **45**, 60–7.
18. Abell, D. F. (1980). *Defining the Business: The Starting Point of Strategic Planning*, Prentice Hall, Englewood Cliffs New Jersey.
19. Lunt, S. T. (1975). 'Trade cycling in U.K. industry'. *Measurement and Control*, **April**, **8**, 152–6.
20. Brittan, S. (1983). 'The myth of the Kondratieff'. *Financial Times*, **7 April**, 23.
21. Brix, V. H. (1967). *Cybernetics and Everyday Affairs*. London, David Rendel, p.144.
22. Kondratieff, N. D. (1926). 'The long waves in economic life'. Translation (1935) in *Review of Economic Statistics*, November.
23. Smith, F. (1981). *The Outlook for the Eighties*. South Africa. 8th National Symposium of the Oil and Colour Chemists' Association. Durban. *Journal of the Oil and Colour Chemists' Association*, **64**.
24. Ray, G. F. (1979). *Some Economic Aspects of Innovation*. Symposium on 'Innovation Studies in the U.K.' Polytechnic of Central London, p.25.
25. Simiand, F. (1932). *Le salaire, l'évolution sociale et la monnaie*, Paris.
26. Schumpeter, J. A. (1939). *Business Cycles: A Theoretical Historical and Statistical Analysis of the Capitalist Process*. London, McGraw-Hill.
27. Weston, M. (1983). *The Engineer and the Acceleration of Technological Change*. 27th Graham Clark Lecture, Council of Engineering Institutions, London.
28. Mensch, G. (1978). '1984: a new push of basic innovations?'. *Research Policy*, **7**, 108–22.
29. Forrester, J. W. (1979). 'Innovation and the economic long wave'. *The McKinsey Quarterly*, **Spring**, 26–38.

30. Chambers, J. C., Mullick, S. K., and Smith, D. D. (1971). 'How to choose the right forecasting techniques'. *Harvard Business Review*, **July–August**, 45–74.

4

Growth and communications

4.1 THE DEPENDENCE OF COMPANY GROWTH ON INFORMATION

Strategic planning must take account both of competitors' activities and those environmental factors which are thought likely to influence the future course of a business. Reliable information on the environment and the competition is, therefore, a key prerequisite in the growth process. To be useful, however, the information must first be understandable, which implies that there will be an instant recognition of its affinity with, and relevance to, what is already known. It must also be restricted in quantity, otherwise it will either swamp the system with trivialities, or be inadequately considered due to lack of time.

The transfer of information is not easy; its value is not absolute but is time dependent. Pearson(1) has pointed out that the value of information is also affected by the personal characteristics of the recipient, the state of activity in which his company is involved, and that nature of the organization in which he is positioned.

Commercial information of a quality needed to make sound policy decisions, whether qualitative or quantitative, is difficult to acquire. It is often incomplete, inaccurate, or out of date, and is consequently associated with so many uncertainties that it is nearly always necessary to apply judgements in order to decide which options to adopt in order to gain a stated business objective. It cannot be easily presented in a compact, and completely dependable form. Finniston(2) has commented that the two formal methods which he uses for discovering what a competitor is doing are Extel, and annual reports and accounts, both of which are somewhat outdated. He remarks that the latter does not supply information on the large grey areas between the turnover and the trading profit. The basic problem is that the questions with which commercial information is concerned are difficult to structure, and there is, as yet, no body of knowledge as in the sciences, and no immutable laws to which reference can be made.

Dare(3), in discussing information required by managers for decision-making in business, has observed that there are no business indices and abstracting services equivalent to *Chemical Abstracts* and *Engineering Index*. Other problems encountered when seeking commercial information are the multiplicity of available services, lack of coordination between sources of widely varying competence, regional bias, under utilization of statistics completed by government mainly for their own use, and a dearth of information from foreign sources.

The provision of financial data is an area in which noteworthy progress has been made. Information is converted into digital form, entered on a database by computer techniques, and is electronically delivered to a customer's interactive video system which comprises a view data screen and a hard copyprinter. There is great variation with respect to the kind of information stored and how it is relayed to clients. Suppliers may ask for a profile of clients' information needs in order to ensure that data is designed in the most appropriate form. The client's desktop terminal may also be used to carry out analyses on the data held by the supplier. Reuters has applied these principles with success, and their provision of specialist finance information gained them 17 per cent of the market share of the 1982 European revenue for databased service, and a record pretax profit of £36m (See Snoddy(4)).

Turning to the component of growth with which this book is mainly concerned, in other words, the development of new products, the process normally begins with the formulation of a new concept whose value will reflect the store of the inventor's knowledge and his ability to augment it. In all subsequent stages there will be a requirement for data and facts, and where these are inadequate success is unlikely. Rothwell(5) has analyzed 14 of failure in the textile machinery and scientific instrument industry, and has shown that poor information was a major cause in all cases. His analysis distinguished between internal and external communications and showed that the former contributed to eight failures and the latter to 12, while seven were attributed to staff ignoring outside advice. The study further indicated that with better information at least 50 per cent of the innovations could have been successful, and the cost of failure of the remaining 50 per cent could have been significantly reduced.

Information is clearly an essential ingredient of a business but it has a transitory quality and is not easily communicated. The skill with which it is handled can be important in determining the success of a business. In recent years it has been subject to worldwide scrutiny and its role is beginning to be understood.

4.2 FORMAL COMMUNICATIONS WITHIN A COMPANY

A prime responsibility of top management is to see that employees are not only informed of company goals, and both strategic and operating plans, but also be given a summary of the company's financial results. The regular supply of information can be a crucial element in the maintenance of high morale. Particulars of competitors' performance should also be circulated, because many employees are strongly motivated by a sense of urgency derived from a determination to achieve and maintain a leading position in the market.

In a large company it is inadvisable to use a common method of communication, but rather messages should be targeted at specific groups of employees and based upon an understanding of their problems and needs. The electrical analogy is apposite wherein communication cannot take place unless receivers and transmitters share a common wavelength.

Managers should set up communication channels which will enable information to flow throughout an organization, and particularly across functional boundaries. In development work the channels will follow paths indicated by the organizational structure and few difficulties will normally arise, but, as discussed below, information in a research department will flow in a more informal way. When members of project groups are assigned from different divisions and departments communications are usually good, but subgroups should also be encouraged for, even if they only meet occasionally, a greater awareness of the business will be generated at the working level.

4.3 FORMAL COMMUNICATION CHANNELS—
LIBRARIES AND OTHER ORGANIZATIONS

A brochure, issued by the British Library in 1982, states that the Library provides a computer search service, and the Lending Division can provide expertise in online searching over a wide range of subjects and files. This extract reflects the revolution which is taking place in information technology and which has profound implications on the way in which libraries and information services are likely to develop in the future.

The advances, stimulated by new technology, have been accelerated by three trends. The first is that the rate of increase of paper documents is so great that doubt has been expressed(6) on the practicability of maintaining individual collections in libraries beyond AD 2000. In illustration of this we note that the 20 million existing patents are being added to at the rate of three-quarters of a million per annum. The second trend is that industry is demanding more information of greater depth and breadth, and, whereas demands were formerly largely centred upon technologies, it is now extending to financial planning, legal, marketing, and regulatory information. The third disturbance has been caused by a rapid increase in the number of information services. The number of online bases vividly reflects this tendency, for, whereas in 1979 there were 400 databanks compiled by 221 producers, in 1983 these figures had swollen to 1 878 and 927 respectively.

Pondering on this rapidly changing scene, Laver(7) concluded that the purpose of libraries is not to hold and issue documents, but rather to make known and available the contents of their collections. This function will be made easier as information technology advances the science of using computers, word processors and telecommunications, to store, retrieve, despatch and receive information. Greater availability and decreased cost of online terminals, in conjunction with the increasing number of sources, opens industry to a vast store of previously untapped information.

The success of the new development must be judged by users who will assess the ease and speed of access to information, range and completeness of coverage, simplicity of use of the systems, reliability and cost, and overall effectiveness. In order to appreciate the impact which the new information technology is having on information services it is instructive to consider the pharmaceutical industry, since this sector, because of the complexity of their problems, is unique in that its staff resources were greater in 1983 than in 1977(8). The major stages in the discovery and development of a successful new pharmaceutical drug involve no less than 15 disciplines(9) and a vast amount of data must be generated and collated during product development in order to provide information to the various regulatory bodies around the world before a drug can be tested on human beings, and, if satisfactory, later marketed for a defined therapeutic use.

Ward(9) describes four objectives of the Glaxo information services: to make certain that the company exploits relevant published information, to collect internal information of long-term importance, to disseminate and organize for retrospective retrieval, and to act as a source of technical expertise. This service is supported by comprehensive computer, terminal and word processor facilities, while, by means of an extensive subject indexing programme, a library book database is available directly online to laboratory staff.

It is interesting to read in Ward's paper that graduates from the information department are assigned to research development and product groups, and assume responsibility for developing services specifically for the group in which they are placed. The department adopts both a service and strategic role, and integrates the functions of the library, the information officers, and the industrial decision-makers.

There is a widening recognition that because the gap between the professional user of information services and his casual counterpart is becoming larger, there is an increasing need for information scientists. It is also interesting to note that Ward's department is divided into three sections that deal with library, literature, and data and document services, respectively. In a small industrial company librarians may have to combine all three functions. They will need to give help with such matters as searches on computer databanks, since an inexperienced user could neither decide which databanks to interrogate nor know how to structure a query in order to obtain a manageable response. They will also probably need to play a more interpretive role and help users with pertinent and precise formulation of their requirements.

Information technology is changing too rapidly to justify detailed descriptions of current developments, but a signpost to the probable future is indicated by Ward's summary of future developments:
Library
    Complete automation
    Training of end-users
Literature services
    Removal of clerical effort from online retrieval
    Improvement of selective dissemination of information
    Training of end-users
Data and documentation services
    Graphs access
    Direct data capture
    Direct scientific use

4.4 COMMUNICATIONS IN A R & D LABORATORY

Irrespective of the availability or quality of external sources of information both the background and training of staff condition their approach to the seeking and use of information. To illustrate this observation Parker(10) made an analysis of an inhouse journal circulation list. Table 4.1 shows the distribution, according to qualification and status, of the most popular 25 journals distributed among R & D staff.

The population is too small to treat statistically, but the data confirms what had been generally observed, namely that university-trained scientists and engineers are seen to make more use of books, journals and abstracts than do non-degree staff and are, indeed, responsible for the majority of requests for library additions. They are to the fore in attending lectures and conferences, and are characterized by their habit of keeping personal files on subjects in which they are especially interested.

The non-university degree man not only reads less often and less widely, but attends fewer lectures and conferences. Information is sought from trade catalogues, from staff of associated companies, and from suppliers and manufacturers. These queries, however, are generally quite specific and are rarely intended to evoke help for solving a major design or technical problem.

Table 4.1   Number of journals on circulation list of staff members

On circulation list	Number of staff	University	Postgraduate degree	Senior management	Departmental manager or assistant	Technical qualification only
Of twelve to four journals	17	9	7[1]	7[2]	9	3
Of four to one journals	16	3	0	3	13	12

Note: [1] Four senior managers have postgraduate degrees

        [2] Six senior managers with university education

These two contrasting habits for seeking information are reflected by the fact that, with few exceptions, the university-trained men are alone in not having to be urged to write up their work regularly. A contributory factor, however, may be that a significant proportion of university-trained staff carry out the basic research in which regular stocktaking is a part of their thinking process.

Table 4.2, taken from a paper by Calder(11), summarizes the results of a questionnaire which he issued to 1 082 technicians in the electrical and electronic industries. It seems that a span of 25 years has seen little change in the reading habits of technical staff, which were even then related to their professional training.

Table 4.2   The relationship of professional training to reading habits

Professional training	Per cent of sample	Mean number of journals seen regularly
Higher degree	2	11.5
First degree	15	6.1
Technical qualification only	22	5.8
No qualification	61	3.7

If the approach to seeking facts from outside can vary so widely within one laboratory, even more marked differences may be expected among firms and between countries, each with their own culture and educational systems.

Experience in training groups for problem solving has shown that research and development staff do not always welcome exchange of information. Scientists with postgraduate degrees, and university-trained engineers, have said that their greatest satisfaction comes from solving problems unaided. The supply of information was likened to receiving, unsolicited, the solution to a crossword clue. Others have considered that to seek assistance is a confession of inadequacy, while a few refuse to disseminate information for reasons of self-aggrandisement.

Where there are practical problems requiring urgent solutions it might be thought that reticence would vanish, but personal experience indicates otherwise. Some years ago an information retrieval system was developed to hasten progress when formulating and devising the best production techniques for new brake linings. The concept was to record current experience in such a way that relevant and accumulated past experience could be quickly and easily retrieved to solve, or partially solve, new problems. Technical staff agreed that this idea was good and practical and began by feeding in appropriate information. However, long before the system could be expected to be useful (six to 12 months) there was a reluctance to continue, and the method had to be abandoned.

A physicist in a senior staff position, who had made a practice of circulating to his colleagues a one- or two- yearly review of the literature on friction and wear, extended the voluntary service by issuing a review of some ten years' laboratory reports on formulation of friction materials together with a number of derived hypotheses. Enquiries after two months showed that the majority of staff most likely to profit from the work had not read the report, and showed little enthusiasm for convening groups to discuss its various aspects.

Confirmation of this 'reluctance syndrome' came from the laboratory's information officer who, when asked what he considered would be his most worthwhile future achievement, replied 'the establishing of more frequent and articulate communication within the R & D division.' He added that staff did not freely divulge information and contact with colleagues was avoided, even when there sometimes appeared to be a solution to a problem in a neighbouring room. Not only is there reluctance to accept information, but an examination of 20 successful annual reports on an R & D division, revealed a disinclination to use information even when it was obtained. Descriptions of a number of management tools and new technologies, ranging from statistical planning of experiments to group technology, were judged worthy of trial but were not taken up. At a later date, however, interest was rekindled by chance meetings with enthusiastic practitioners at conferences, seminars, and lectures, and then action quickly followed. In a few instances trials failed because insufficient time had been allowed for the supplier to overcome teething troubles. It was noticed that, once rejected, there was a reluctance to repeat a trial, and two or three excellent opportunities were thereby lost.

An R & D director's vigilance must, therefore, extend to the use as well as to the provision of information.

The above studies are illustrative of the numerous difficulties that hinder the transfer of technology. Behavioural patterns, similar to the above, may also be applicable to production, planning, marketing and, indeed, to the majority of company functions.

Experience gained from working within a wide variety of companies has shown that small and large firms exhibit marked differences in their use of information. The former are not, in general, able to make full use of information sources, and tend to be shortsighted and introspective; the latter fail to give sufficient consideration to the status of information, and so fail to establish its lines of communication paths across all company functions (see also Yates(12)).

4.5 DIFFUSION OF INNOVATIONS

A topic which may possibly be of greater future importance is the process by which ideas are transmitted and developed within a population. Goffman and Newill(13) have treated the subject mathematically and, on the assumption that it is analogous to the spread of an infectious disease, have constructed a model which can be used to answer questions which include (1.) Where and when is an activity within a given discipline developing into epidemic proportions; and (2.) What is its expected duration? Coadic(14) has more recently confirmed Goffman's work, and concluded that the diffusion of ideas is a dynamic and orderly social system exhibiting regularity and law. He analysed the reception of new ideas by researchers in four selected disciplines, and noticed that their variation with time followed definite but different patterns. The quickest diffusion rate was in technology, and the slowest in the social sciences. Figure 4.1 shows that for the former a new idea reached 20 000 individuals in 22 months, and for the latter the corresponding figures were 9 000, and 46 months. Where Goffman's data referred only to large formal communication systems, Coadic added an estimate of the small, but flexible oral contribution.

These studies should benefit information scientists and marketing staff who need to assess the results of introducing new

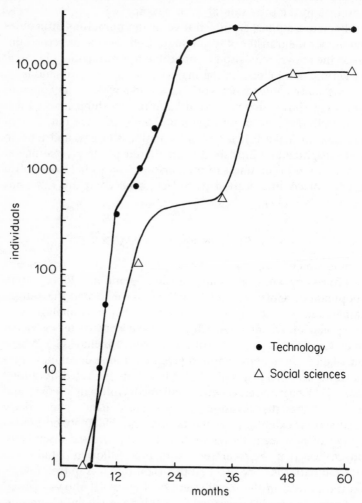

Figure 4.1 Scientific information flow—comparative study

methods of advertising. The work may also have a role in technical forecasting. For example, an industry which is questioning
whether or not it is approaching maturity could construct the
diffusion rate graph for ideas considered to be important to its
technical future, and note whether or not the curve is tending to
become parallel to the x-axis.

The above studies are concerned with the transmission of ideas within a large population and have many practical implications, but an innovator, wishing to sell a new product, will be more concerned to know which factors affect the adoption of new ideas. The classic work on diffusion in innovation is described in a book of this name by Rogers(15) but more recently Hayward and associates(16,17) have described how the application of the Rogers approach was applied to the marketing of capital equipment in the United Kingdom industrial field. He found that diffusion was slow when the innovations were outside the normal practice and understanding of the would-be adopter. Submissions which were merely improvements on an existing practice achieved a 50 per cent adoption rate within approximately ten years, whereas technical breakthroughs needed 20 to achieve the same result. A field study by Parkinson(18) indicated that companies who were quick to adopt innovations were those which used up-to-date management tools, were themselves good at innovation, and skilled at gathering, storing, and using information. They, moreover, exhibited entrepreneurial characteristics in that they not only clearly perceived the advantage of innovation but were also not unduly deterred by the perceived risk. However, in general, the less the perceived risk the greater is the diffusion rate, and this underlines the importance of trying to lessen uncertainties. The seller should supply the would-be customer with a sufficient quantity of information of the highest quality. The information must not only dispel all possible areas of doubt but the sources must be clearly creditable. It must, additionally, be matched to the ability of the recipient to comprehend its nuances. Although in the initial stage information may necessarily have to be impersonal, an early opportunity should be taken to establish more personal contacts. For example, the introduction of a satisfied customer to a prospective buyer can often be a deciding factor.

4.6 INFORMAL COMMUNICATION CHANNELS—
INTERNAL AND EXTERNAL SOURCES

The informal gathering of information is perhaps more interesting, and no less important, than formal channels, and there can be few discerning directors of research and development who have not noticed that crucial facts seem to have been acquired through chance. Czepiel(19) has described the essential step in the information diffusion process as the way an idea, known to one individual, is passed on and used by a second. The process, he believes, is controlled by their social relationships. The source of facts which undergo informal circulation has attracted many researchers, and a review by Rothwell(20) shows that over half of project initiation ideas come from within an organization. There is, additionally, an effect of organization size, and for firms with over 1000 employees external sources predominate, whereas when the number of employees is less than 200 70 per cent of the ideas originate inhouse.

The important questions are: through which channels does an individual discover new facts, and do external and inhouse ideas travel along different routes? Work in this field has been stimulated by a paper 'Communication networks in R & D laboratories' which Allen published in 1971(21). The methodology used was to select pairs of companies which had been awarded identical US government contracts, each of which worked independently to achieve identical project objectives. This made it possible to establish criteria of performance, because comparisons could be made of contract completion time, costs, the quality of the products, and other factors. Firms were rated either as high or low performers. Many variables were examined to account for performance, but the outstanding one was the level of interpersonal communications.

The work showed that the use of information obtained from colleagues within the organization was correlated with performance. More time was spent by high performers in consultation with staff, both within their own and other specialized fields. Furthermore, the high performers gained a lot of information from inhouse sources, but even when they went outside the company the information was also of high quality. The low performers gained more from outside than in, but the quality was poor. It was estimated that over 80 per cent of ideas came from talking to other people, and less than 10 per cent direct from the literature.

Allen was concerned to explain the paradox that whereas successful products were based on many ideas obtained from external sources, the crucial factor for success was the circulation of internal sources of information. By charting sociometric maps of communication patterns in the laboratories, it was observed that the

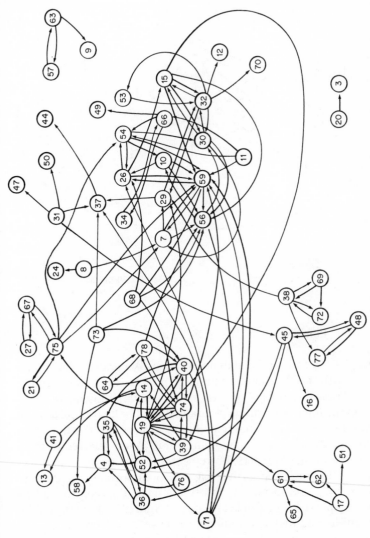

Figure 4.2 Typical communication network of a functional department in a large R & D laboratory

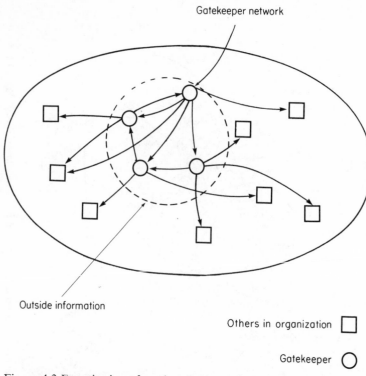

Figure 4.3 Functioning of gatekeeper network

workplace of four or five individuals in every hundred was coincidental with the termination of a large number of communication lines. These points were termed 'nodes' and are illustrated in Fig. 4.2. In the high-performing laboratories staff who occupied nodal positions had an outstanding ability to acquire information from both outside and inside the organization. Allen referred to these individuals as gatekeepers, or key communication stars, and recognized that their supportive role was an important ingredient of a laboratory's performance. Information was thus controlled by a two-step process; the gatekeepers first acquired and monitored information, and then distributed it selectively to the relevant parts of their departments (Fig. 4.3). Although they acted as internal consultants, this was not done consciously, and gatekeepers were unaware of their important contribution to success. The role was conferred upon them spontaneously by their colleagues and was independent of management action. Gatekeepers were generally high performers but did not necessarily occupy management positions. They had a number of common traits. They had strong hobbies, were very active, were twice as likely as the next person to have a postgraduate degree, were publishers of many papers, were neither particularly introvert nor extrovert, and had a strong sense of purpose.

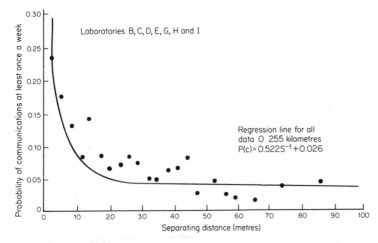

Figure 4.4 The probability that two people will communicate as a function of the separating distance

Arising from studies of the sociometric maps Allen found that information flowed more freely when the places of work of potential communicators were near to each other. Figure 4.4 illustrates this finding, namely, that the probability of communication falls off inversely as the square of the separating distance and is negligible for distances greater than 30 metres. When an organizational affiliation is present the distance effect is still operative but the relation moves to a higher curve. There is a higher probability that an individual will walk a given distance to talk to someone in his group than to someone in a different group (Fig. 4.5).

The author has, on several occasions, experienced the contrasting communication pattern caused by either close proximity or wide separation. The first arose when the new appointment of

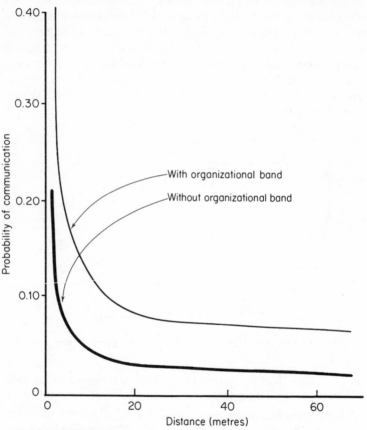

Figure 4.5 Probability of communication with and without an organizational bond

an exceptionally talented individual to a senior position failed to give the expected high performance. On studying Allen's work it was decided to move the individual from his office, remotely situated in a three-pronged building, to a central position adjacent to his colleagues at the same management level. There was a rapid and spectacular improvement. On another occasion, when a laboratory was moved, temporarily, some 730m from the test house which it served, staff visits between the buildings largely ceased and work suffered.

Hough(22) carried out an investigation into a means for improving the flow of technical information between an applied research and a headquarters' technical services laboratory, in more than 50 major overseas markets in which some customers were 16 000 Km distant. Projects ranged from the simplest technical problems to completely new applications, and failures, when they occurred, were not attributable to technical shortcomings but rather to delays that were involved when transmitting information. The level of communication, which was expressed by the total project costs for a given period divided by total market sales, decayed with distance to give a curve similar in shape to that found by Allen for spatial separations of a few hundred metres. The level of communication fell by a factor of ten in 3200 Km, and thereafter remained constant up to the largest distance at which data was obtained, namely, 12 000 Km.

Two mechanisms were adopted to shorten the technical link: the provision of a computer service for which the input project characteristics could be transmitted by telex, and the setting up of local laboratories capable of dealing with 90 per cent of the queries. The remaining 10 per cent were referred to headquarters when fruitful dialogues were established between the two parties.

Allen extended his work concerned with separation of staff within a building to separation between buildings spaced by distances of up to 240 Km apart. There was no clear communication pattern unless there was some organizational bond between the staff. The communication was fair when one department shared two buildings, and good when staff shared a common discipline. Work in Japan has shown that communication between technical staff in two separated subsidiaries of a company improved when gatekeepers in the two places changed places. The improvement, however, diminished with time in an exponential manner and became negligible in two to three years. When information is required to flow along informal channels between separated buildings it is important to adopt one of the above or similar methods, for Allen has observed that, with a fall in personal communications, goes an associated decline in telephone conversations and written communications.

Researches on interpersonal communications have been reviewed by Epton(23) who observes that many open questions remain. There are also several points of contention concerning Allen's statement of causality. The doubt is whether good communication produces high performance or whether the good performer appreciates the necessity of better communication. It may be that some confusion arises because authors have failed to describe in sufficient detail the nature of tasks and the characteristics of staff in the various organizations under investigation. In his book published in 1977 Allen(24) clearly distinguishes between scientists who, working on highly specialized subjects, associate in 'invisible colleges' and keep each other informed of their work and technical engineers who keep abreast of information by associating with colleagues in their own organization. In his 1970 paper, however, the investigations refer to R & D laboratories staffed by engineers. Because of cultural differences between countries, readers in the United Kingdom tend to accept that all laboratories, even those in the United States, are staffed by scientists unless the contrary is stated. These, and other misunderstandings, may partly explain discrepancies noted between results obtained in the two countries.

Studies on informal communications contain important implications for those companies wishing to innovate. Managers should observe which members of staff appear to fill the role of gatekeeper, and place them in offices that are strategically sited in order best to meet the operational needs of projects. Existing networks should be strengthened, and interfunctional and interdisciplinary teams should be formed and distributed to legitimize interchange across formal organizational boundaries. The work of Taylor and Utterback(25) showed the converse side of this picture, when they demonstrated that changes to an organization structure which broke bonds of group assignment, technical assignment and location simultaneously, had a most damaging effect on communication. Changes both to organizations and the composition of teams are necessary if creativity is to be satisfied, but they should be done in a manner which will facilitate rebuilding crucial information links.

Two further studies by Allen and collaborators have practical implications. Cooney and Allen(26) have suggested that the gatekeeper phenomena plays an important, perhaps a dominant, role in the transfer of technical information among nations. Scientists' time can be, therefore, more effectively spent in meeting colleagues for discussion than in reading or using computerized information retrieval systems. Job mobility, foreign sabbaticals and fellowships, and other forms of extended foreign technical experience should be recognized by governments, international agencies, and international companies in their programmes for improving technical flow.

Allen and Fusfeld(27) have shown that communication is influenced by the physical and architectural arrangements of a laboratory. Communication between individuals is sensitive both to the horizontal and vertical distance separating them, and, on the assumption that the length of stairs between two floors is at least equivalent to that amount of horizontal separation, a computation can be made of the mean separating distances between occupants in a square building that has a central staircase. The very surprising, and interesting, conclusion is that for optimum communication the laboratory should be a single storey square building for a floor space of less than 10 000 square metres, but above that area the building should have at least three floors and an elevator.

4.7 THE USE OF MODELS IN BUSINESS COMMUNICATION

The average businessman has to deal with an extremely wide range of subjects and will rarely find time for long periods of uninterrupted thought. His strength will lie in his ability to perceive how his organization behaves, and he will apply this knowledge and intuition to improving its effectiveness. Urgency will dictate an essentially pragmatic outlook, and any lack of analysis will be more than made up by energy, drive and skill in making decisions. By contrast, academic staff engaged in business studies will, by concentrating on a narrow range of specialization, be better able to analyse a situation for they will have opportunities to collect relevant data, classify, hypothesize, and predict. Rarely, however, will they be aware of all aspects and nuances of the total up-to-date commercial picture.

Clearly the two situations are complementary and the aim must be to devise ways whereby each practitioner will become increasingly aware of each other's problems and solutions. One approach is to increase comprehension by using visual or mathematical representations, usually referred to as models. With thought

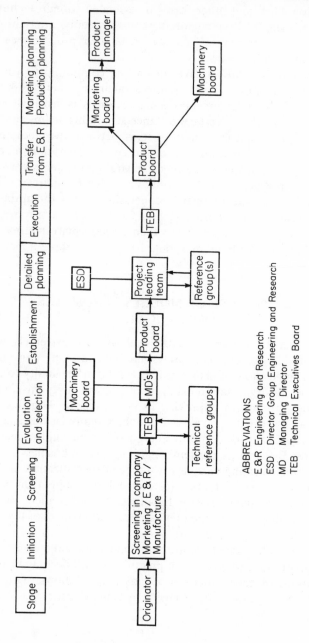

Figure 4.6 The innovative process

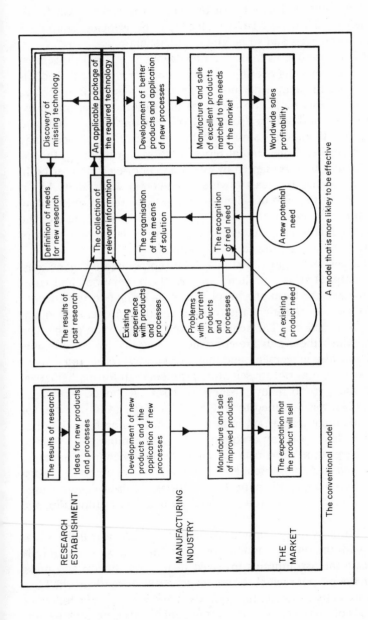

Figure 4.7 A comparison of old and new approaches to the generation of new products and processes

and ingenuity a situation can be depicted in a form which, although it may depart from reality, may nevertheless simplify a complex analysis and so heighten understanding.

It is instructive to look at an analagous situation in the fields of science and engineering. Experiments on small-scale models of cars are made in wind tunnels where designers seek to achieve low wind resistance, and problems of erosion and salting in estuaries and harbours are solved by constructing scale models and subjecting them to the simulated flow of tides. Other models may depart further from factual representations. A building, for example, may be represented by either a plan or elevation, or both. Symbolism has the least obvious connection with reality but it has been at the hub of scientific progress. A recent outstanding example is shown in the use of powerful computers to solve the complex relationships based on the laws of planetary motion in order to guide satelites to their destination.

Business activities can similarly be represented by many aids. The simplest to comprehend are visual presentations and commonly take the form of a graph, diagram, or matrix, the purpose of which is to represent the form of relationship between quantities or functions. The acquisition and manipulation of the data from which these aids are constructed may involve considerable skill and wise judgement, and they must always reflect integrity of purpose. Just as obtuse scientific problems call for mathematical modelling, so do many business analyses. A model, termed Z-score(28), was designed for the Bank of England to provide, for individual companies, an indication of impending financial difficulties. It is based on accounting data, and initial trials showed it to have some value as a screening device, enabling companies to be selected for subsequent conventional examination. As discussed on p.53, mathematical calculations are also frequently used to predict the future growth or decline in a firm's production and sales.

A business model should simulate accurately a real-life situation with the fewest possible complications, and should carefully indicate the assumptions involved in its construction and make clear the purpose which it intends to serve. It can help the business executive to recapture quickly a train of thought following periods of unavoidable interruption, aid his decision-making, and help his colleagues to appreciate the essential elements of an argument or proposition in the shortest possible time.

Figures 4.6 and 4.7(29, 30) show two uses of diagrams related to the innovation process. The first clearly shows the stage at which executive staff, senior managers and departments become actively involved, and, suitably annotated and updated, can serve as a quick reminder of how far a project has progressed. The second illustrates that a reliable definition of needs for a research and development project requires both information from many sources and an exploration of possible technology transfer.

The facility of computers to store and retrieve data and present coloured arrays, is encouraging the use of models of ever-increasing ingenuity.

4.8 SUMMARY

Top management should inform employees of the company's main aspirations and achievements, and communication channels should be set up to ensure that information flows across and between all functional activities. Adequate and reliable information is more difficult to obtain on commercial than on technical and scientific matters.

There is an ever-increasing and widening demand for information, and the functions of libraries are undergoing radical change. Information services, of which libraries may be a part, are now providing both a strategic as well as a service role, and as these become increasingly sophisticated inexperienced users will need guidance from information scientists.

The way in which staff collect and use information is conditioned by their training and education, and an understanding of the differences is essential if full advantage is to be taken of advances in information technology. A major obstacle to the transfer of knowledge among research staff, is a belief that to seek assistance is a confession of inadequacy. New concepts are rarely spread by reading, but rather from contact with enthusiastic practitioners.

The means by which ideas are transmitted and developed within a population is analagous to the spread of an infectious disease. It is a dynamic and orderly social system, and the spread of ideas with time shows regular patterns whose form varies among different disciplines.

Attention paid to those local factors which control the adoption of new ideas shows that diffusion is slow if an innovation is outside the normal practice and understanding of a would-be adopter. Not surprisingly, entrepreneurial companies have been observed to be more ready to adopt innovation and are not unduly deterred by perceived risks. Suppliers should always provide information of the highest quality in order to dispel possible doubts expressed by the users.

In the research and development process both formal and informal challenge of information are equally important. In an organization which employs over 1000 people, external sources of information predominate, but when there are fewer than 200, ideas mainly originate within a company. Allen's work on information communication networks showed that high, compared to low, performers spend more time in consultation with staff and gain much high quality information, both from within and outside the company.

In high performance laboratories there were staff termed gatekeepers who first acquired and monitored information, and then distributed it selectively to relevant parts of the organization. They were unaware of their internal consultant role which was conferred upon them spontaneously, independent of management action.

Information between the potential communicators flowed more freely if their places of work were close together, and for spatial separations of a few hundred metres Allen found that the probability of communication decreased inversely as a square of the distance of separation. A similar inverse delay law operates between staff housed in widely separated buildings and between a headquarters and overseas staff separated by some thousands of kilometres. It is helpful to form interfunctional and interdisciplinary teams in order to legitimize exchanges between separated locations across formal organization barriers.

Communication can be facilitated by designing models, and they can perform a valuable function in helping both business and academic staff to understand each other's points of view. Models should be designed to simulate real life situations in a simple manner: they help executives to recapture a lost train of thought, facilitate an understanding of a new concept, and so speed decision-making.

4.9 REFERENCES

1. Pearson, A. W. (1973). *Fundamental problems of information transfer*, ASLIB Proceedings. **25**, No. 11. 415–24.
2. Finniston, M. (1984). *Information for Industry: The Next Ten Years*, British Library R & D Report 5802, **April** 43–6.
3. Dare, G. (1984). *Information for Industry: The Next Ten Years*, British Library R & D Report 5802, **April**. 22–32.
4. Snoddy, R. (1983). 'Data base services may grow by 20 per cent'. *Financial Times*, **25 October**, xxvi.
5. Rothwell, R. (1984). *Information for Industry: The Next Ten Years*. British Library R & D Report 5802, **April**, 8–14.
6. Lester, R. (1984). *Information for Industry: The Next Ten Years*, British Library R & D Report 5802, **April**. 63–73.
7. Laver, M. (1983). *Information, technology and libraries*. The First British Library Annual Research Lecture, The British Library.
8. Jones, A. (1984). *Information Demand and Supply in British Industry. 1977–1983, Information for Industry: The Next Ten Years*, British Library R & D Report 5802, **April**. 15–21.
9. Ward, S. (1984). *Pharmaceutical Industry Information Services and the Impact of New Technology. 'Information for Industry: The Next Ten Years*, The British Library R & D Report 5802, **April**. 47–62.
10. Parker, R. C. (1975). *Communication Patterns in an Industrial R & D Laboratory*. British Library Research & Development Report, No.5255, 53–59.
11. Calder, N. (1959). *What they Read and Why*. DSIR, Problems of Progress in Industry. HMSO.
12. Yates, B. (1973). The place of the information service within the organizational structure. *ASLIB Proceedings*, **25**, No.11, 430–4.
13. Goffman, W. and Newill, V. A. (1964). 'Generalization of epidemic theory. An application to the transmission of ideas'. *Nature*, **204**, No.4953, 225–8.
14. Coadic, Le Y. F. (1974). 'Information systems and the spread of scientific ideas'. *R & D Management*, **4**, No.2, 97–111.
15. Rogers, E. M. (1962). *Diffusion of Innovation*. Free Press of Glencoe, New York.
16. Hayward, G. (1980). 'The adoption of technological innovations'. *Journal of the Institute of Engineers and Technicians*, **Spring**, 6–10.
17. Hayward, G., Allen, D. H. and Masterson, J. (1976). 'Characteristics and diffusion of technological innovations'. *R & D Management*, **7**, No.1, 15–24.
18. Parkinson, S. T. (1975). *The Role of Information in the Diffusion of Industrial Innovation*. Marketing Communications Workshop,

Brussels. University of Strathclyde—Department of Marketing. 1–66.

19. Czepiel, J. A. (1977). *Communications Networks and Innovation in Industrial Communities.* Symposium on Innovation. University of Strathclyde, Glasgow, September, 1–20.

20. Rothwell, R. (1975). 'Patterns of information flow during the innovation process'. *ASLIB Proceedings,* **27**, 217–26.

21. Allen, T. J. (1970). 'Communication networks in R & D Laboratories'. *R & D Management,* **1**, No.1, 14–21.

22. Hough, E. A. (1972). 'Communication of technical information between oversea markets and head office laboratories'. *R & D Management,* **3**, No.1, 1–5.

23. Epton, S. R. (1981). 'Ten years of R & D management - some major themes: the role of communication in R & D Management. *R & D Management,* **11**, 4, 165–70.

24. Allen, T. J. (1978). *Managing the Flow of Technology: Technology Transfer and the Dissemination of Technological Information within the R & D Organization.* The MIT Press, Cambridge, Massachusetts.

25. Taylor, R. L. and Utterback, J. M. (1975). 'A longitudinal study of communication in research: technical and managerial influences'. *IEEE Transactions of Engineering Management.* **EM-22**, No.2, 80–7.

26. Cooney, S. and Allen, T. J. (1974). 'The technological gatekeeper and policies for national and international transfer of information'. *R & D Management,* **5**, No.1, 29–33.

27. Allen, T. J. and Fusfeld, A. R. (1975). 'Research laboratory architecture and the structuring of communications'. *R & D Management,* **5**, No.2, 153–64.

28. Anon. (1982). 'Techniques for assessing corporate financial strength'. *Bank of England Quarterly Bulletin,* **June**, 221–3.

29. Saren, M. A. (1984). 'A classification and review of models of the intra-firm innovation process'. *R & D Management,* **14**, No.1, 11–24.

30. Neale, M. J. (1984). 'Technology for industry—a new initiative by the I.Mech.E.', *Chartered Mechanical Engineer,* **31**, No.2, 51–3.

5

Sources of New Products

5.1 AVENUES TO GROWTH

The manner in which a small business grows largely reflects its internal climate (see Chapter 2) but, as time passes, it becomes essential to pay more and more attention to what is happening outside the company (see Chapter 3). Though the environment is always likely to contain dangers, the degree of uncertainty can be lessened through an understanding of the way in which information passes along both formal and informal channels (see Chapter 4). Eventually, however, there comes a time when the subject of crucial concern is the identification of possible sources of new products, and it is with this that this chapter is largely concerned.

It may be argued that, provided a business is founded upon a product, or products, which satisfy a universal need, long-term growth can be achieved by selling in more and more countries. For this to be possible, not only must a product have intrinsic merit, but be sold at a cost which can be kept competitive by directing continuing attention to the manufacturing process. The first aim will be to establish economies of scale, and this process can be helped if there is a convergence of customers' needs throughout the world. It is often advantageous for each continent to have its own autonomous arrangements for production, sales, and marketing, although to do this it may be advisable to set up a headquarters international division. Top executives also should be experienced in, and knowledgeable about, international affairs.

Coca Cola is a company which fulfils closely the above requirements. An article(1) published in 1979 stated that the product was sold in 135 countries, and, for the decade ending 1979, enjoyed a 12–13 per cent annual rate of growth for both turnover and profits. Although at this time sales of Coca Cola, fruit juices, coffee, tea, and wine totalled $5billion, the 1981–83 business plan contained ideas for diversification units in each of the major countries(2) in which they were established. This example of growth based upon a one-product line is uncommon.

The usual expansion modes will include one or more of the following: a higher usage rate by existing customers, the sale of the same basic product to new customers (including those gained from competitors) in the same or similar market, and entry into new markets. Usage can be increased by trading-in a lower product life for a sought-after improvement in either a product attribute or service, and by modifying the product in order to encourage more frequent consumption. New customers may often be gained by offering an improved product where and when required, at an acceptable price and with tangible and intangible support. The former could, for example, be a specially designed point-of-sale aid to help a retailer's promotion, and the latter could arise from establishing very friendly relationships with factors. Entry into new markets could be based upon exporting existing products, but most businesses will need to expand their product portfolio.

All but one of the above five subdivisions call for new or improved products, and, in Chapter 7, it is suggested how an ongoing search for new products can be set up, while Chapter 9 deals with their assessment and selection. Although many advantages accrue from seeking ideas within a company, an optimum choice cannot be made without an examination of all external sources of new products. This chapter, and the following, deal both with possible external sources and the incorporation of ideas within a business plan.

The number of product types tends to increase with company growth and may reach a level at which it is suddenly realized that product complexity is threatening both good communications and sound decision making. To prevent this happening a model should be constructed to differentiate between three classifications of products, namely, lines, forms and types: lines refer to a group of products which are closely related and satisfy one need; the form is a subdivision and signifies a common design principle; while the further subdivision to types represents the application of a design principle to a particular practical application. The model will show whether existing product lines fit the business definition, and will facilitate identification of gaps in the product portfolio. If new additions are indicated, the model can give an early indication of necessary organizational changes.

Figures 5.1 and 5.2 illustrate the possible range of products for: (a) a business concerned with transmitting and also converting energy into its lowest form, and (b) the manufacture of electronic calculators.

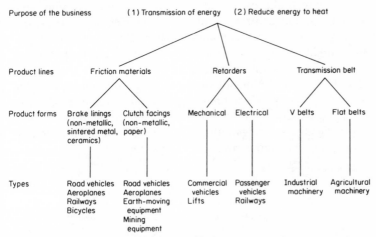

Figure 5.1 A model for products supplied to manufacturers and consumers

In the first example the products are mainly component parts for use by other manufacturers although most types were also marketed for direct retail sales in the replacement market, whereas the second example solely concerns consumer products. The two examples are chosen to demonstrate that even for dissimilar business structures a model can be drawn to show that both comprise several product lines each of which has more than one form, and that these in turn have a number of types. The contrasting characteristics of products within each grouping, and hence development requirements, decrease with the process of subdivision to types. In large organizations the product lines may require their own divisions or subsidiaries, while product forms may also require separate marketing and possibly separate technical departments.

Figure 5.1 shows clearly that the potential for company growth is related to the way in which the purpose of the business is expressed. The more the definition is restricted to general principles the greater will be the number of possible product lines.

The model, suitably expanded, could serve a number of purposes. For example, motor vehicles could be subdivided into motorcycles, cars, and commercial vehicles, and if technical information was added, an indication could be obtained of the extent to which disc brake pads, and brake-drum linings, would need modifying to suit contrasting duties of vehicles sold throughout the world. Again, a comparison of usage between cars and aeroplanes might be thought sufficiently great to merit specialized sales forces and technical groups.

To obtain new business, consideration would first be given to introducing new product lines. Should, for example, abrasive wheels and papers be thought to fit the business definition of Fig. 5.1, preliminary investigations would be made of the technology, market potential, and competitors' activities.

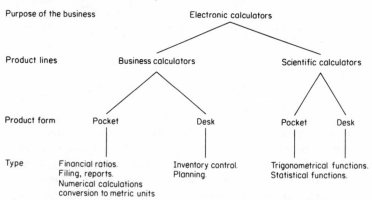

Figure 5.2 A model for products supplied to consumers

Figure 5.2 indicates how a more restricted business definition limits the scope and hence complexity of product lines, forms, and types. Both business and scientific calculators call for similar technical and market expertise, and this would be reflected in a relatively simple organization structure closely paralleling the product diagram. New product lines, though more difficult to find in this somewhat circumscribed field, would be likely to fit the existing technology and marketing structure.

As we have seen earlier, competitive new products may result from the evolution of current designs and manufacturing techniques, but when the potential for further evolution is exhausted innovation must be introduced. It is noteworthy that the outstandingly good growth company 3M adopts a criterion that 25 per cent of the product line must be new within the last five years(3).

The range of technical expertise needed to deal with the product lines shown in Fig. 5.1 demonstrates the dangers inherent in placing too much reliance upon inhouse activities. Indeed, the failure of many medium-sized United Kingdom companies during the recession can be attributed to a rejection of possible new product lines for the reason that inhouse research and development facilities were inadequate, and a belief that external sources for diversification lay only within the province of large companies. When drawing up business plans, and discussing strategies, it is imperative to carry out a thorough survey of every possible source for new product lines.

5.2 THE INITIAL SEARCH FOR NEW PRODUCTS

5.2.1 General

Realization that a company needs new products may be triggered off by a wide variety of circumstances, and there are many imperatives to which a company should be sensitive. If competition is fierce it may be necessary to adopt a new technology in order radically to improve an existing product. A new product in a different, but related, technology may also be needed to help create a potential synergistic improvement. The product range may also need supplementing in order further to satisfy existing customers and so to render less likely the entry of competitors. Expansion into a particular market segment, or into another country, may require a new distribution chain and other organizational changes. Information may be gained that an important manufacturer, with whom a competitor has enjoyed sole supplier status, in adopting a policy of two or more equal suppliers, is to offer opportunities for additions to the company's product range. A company may also be forced to avert threats to a vital raw material.

Because the need for new products appears in so many guises it follows that there may be wide variations in time scales, patterns of risks, and demands on resources. The usual path lies in inhouse development but many other approaches must be considered. The needs will frequently be urgent and require additional resources; it will then be necessary to review means of either augmenting or reallocating resources through possible divestment of company divisions, or subsidiaries, or the abandonment of product lines, forms, or types.

There are nine avenues which can lead to new products, and a company may decide on several approaches at any one time. The first group of four is illustrated in Fig. 5.3a and comprises inhouse research, development, and design together with three methods of augmenting this activity, namely, licensing, technical interchange and subcontracting. Though the latter two may involve realignment and, possibly, reorganization of research, development and design staff, significant restructuring of the company will only be necessary when input of new technology causes dramatic changes to either, or both, manufacturing and marketing.

The second group is characterized by rearranging discrete activities and comprises mergers, conglomerates, joint ventures, acquisitions, management buyouts, and divestments. The first, second, and third are aimed at growth and, as Figure 5.3b indicates, involve a considerable degree of restructuring. On bringing the separate activities together, new and complex relationships will be forged, and because the endpoint will be a new and distinct corporate entity the process must be guided with great perception. Conglomerates do not involve structural changes, and reflect a strategy the aim of which is to avoid synergy. The last two illustrations in Figure 5.3b concern the elimination of business activities, and the purpose should be to release funds which may be used for new, and more profitable products. Activities other than research can be subcontracted and may be a cautious first step towards adopting a more formal path to diversification. For example, Eastman Kodak planned an entry into the 8 mm video market, ahead of the world's 122 manufacturers, by becoming

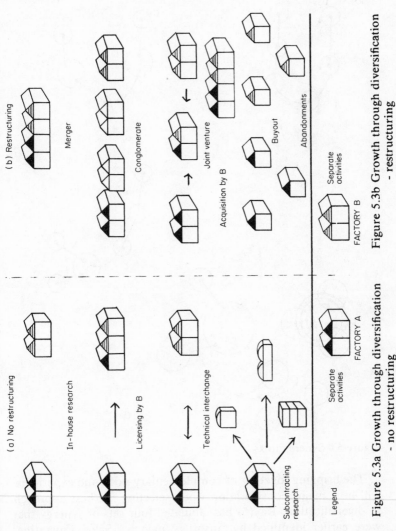

Figure 5.3a Growth through diversification - no restructuring

Figure 5.3b Growth through diversification - restructuring

the market arm of two Japanese companies, Matsushita for video camera systems and TDK for tapes(4).

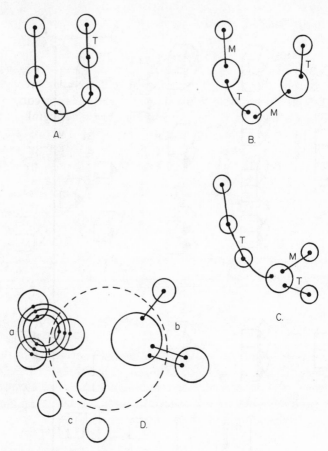

Figure 5.4 Synergy links

The bringing together of complementary skills and experience will be a major consideration when selecting the best approach to diversification. Kay(5) has tabulated four sets of synergy that were earlier identified by Ansoff(6) namely: Sales, Operating, Investment and Management. Some 15 sources are itemised together with a number of unspecified management factors. These sources of mutually supportive activities can be richly rewarding, for experience has shown that the total effect always exceeds the sum of individual acts of synergy.

In his book *The Evolving Firm* Kay recommends that synergy maps be drawn to provide a framework on which to construct business policies. He illustrates the potential of this concept by drawing four synergy maps (Figure 5.4) in which, for simplicity, each group of activities characterizing a synergy type is represented by one link irrespective of the actual number involved. For example, in example A all six companies share a common technological link. In B there are four discontinuous links, and C represents a more complex strategy in which there is a single technical link between four companies and two single links, one market and one technological. The fourth map is drawn to represent competition between three groups of companies.

Group (a) represents three companies which share three synergy links; in (b) one company has one link with an associate and two with another: and in (c) there are no common links for it is a true conglomerate. Because synergy can rarely be fully exploited, especially when the businesses are large, (a) may not fully exhibit the expected advantage over (b).

Bonds between group activities can be dangerous as well as advantageous. For example, an early strategy of the Turner and Newall Group was to build up a network of companies for the purpose of exploiting their mining of asbestos. After over 70 years of growth the mineral was deemed dangerous to health, and the entire group was faced with a common threat. Although this danger was happily foreseen, and companies were acquired whose products were free of asbestos, it was too late to prevent a considerable slimming down of the group activities.

5.2.2 Inhouse research and development

The function of inhouse research and development varies among companies, and is by no means a universal source of new products. In those founded upon craftsmanship and practical experience, research and development activities may be very small or even absent. Products tend to evolve from a knowledge of customers' needs, and the emphasis will be on evolutionary development. In high technological companies, on the other hand, inhouse research will be the main source of new products.

Between these extremes a gradual transition occurs: as the nature of the technology becomes more advanced research will be expected to provide more ideas for new products, and development to be responsible for their practical achievement. Unfortunately, even at these intermediate levels the research process is not as successful as is widely believed, especially when a mature product is challenged by a new technology. Cooper and Schendel(7) investigated 22 such cases of companies faced with a new competitive technology, and found that of 15 who launched major R & D programmes only two succeeded in repelling the threatened competition. A 1979 study by the US National Science Foundation showed that small US firms produced 24 times as many innovations per unit of expenditure as large ones, and many large corporations are seeking ways to motivate a more entrepreneurial behaviour (see Chapter 8 p.196).

5.2.3 Licensing and Technical Interchange

Of the frictional materials product lines shown in Fig. 5.1 four product forms were manufactured under license: sintered metals, industrial clutch facings, ceramic aircraft brake pads, and paper facings for automatic transmissions. All four engineering components were required for vehicles manufactured by United Kingdom subsidiaries of overseas parents. In the period prior to obtaining a licence, business could not be gained until the subsidiary had submitted samples from prospective suppliers to the overseas headquarters for approval, and this process caused months of delay. By contrast, once production of the licensed material was deemed satisfactory by headquarters staff, general approval was given for future releases. A decision to license is made easier if there is a prospect of achieving the status of a preferred supplier; in other circumstances much more thought would have to be given to the product and its manufacturing processes, to the extent of possible market penetration, and a search would have to be carried out for emerging competitive technologies.

Because a licence is an alternative to inhouse activities, it is particularly advantageous should there be a chance of unforeseen, adverse changes in market prospects. This follows from the probability that an option to terminate the licence will be sooner than the time required to recover research and development costs.

It is frequently claimed that the main attraction of a licensing arrangement is the speed with which a new product can be marketed. This is only likely to be so if the licensor and licensee businesses are similar, in which case the licensee is likely to have the necessary capital plant, and sufficient technical competence both for development and production. The required marketing expertise should also not depart too far from that with which the company is familiar. In every prospective licensing arrangement, attention will need to be paid to taxation and currency regulations, production and engineering standards, patents, and legislative constraints.

Particular care should be taken when translating the economics of a process from an overseas company. The availability and cost of raw materials and energy can cause problems. The cost of labour may also be crucial, and failures have been caused by unions' insistence on the employment of highly paid, skilled labour for operations which were carried out in the licensor's factory by unskilled labour.

Formal arrangements for technical interchange arise in many situations. An agreement between a licensor and licensee may provide for both parties to inform each other of all technical advances. Companies acquired by conglomerates may be required to share their technology with other group companies, especially if activities are global. Indeed any two companies may be required to interchange development and production know-how whether they are linked by trademark, distribution or other types of association.

Successful technical interchange is, in practice, extremely difficult to achieve, especially if two countries are involved. The companies will probably have different management structures, and there will be contrasting norms, added to which may be a difficult language barrier which can be compounded through separation by several time zones. A number of multinationals concerned with the operation of their global businesses are studying contrasts in management styles, and George[8] advises managers to take pains to understand both their own culture and the one in which they are to operate. Meaningful communication will be difficult and, as a result, long-term philosophical concepts behind a piece of research will not be appreciated. Experience with running joint research programmes between companies in different countries has shown that they are interrupted by local crises (such as fire-fighting), so that both parties at one time or another become convinced that their partners are not contributing a fair division of the work. Even an attempt to share work by dividing it among priority groups with some projects chosen to be investigated by both firms, and others allocated to either company according to their expertise, did little to lessen the problem. Following several years of experience it was apparent that technical interchange could only be used for the more routine tasks, such as developing analytical methods and devising new test procedures, which could be accurately prescribed.

5.2.4 Subcontracting Research and Development

Opportunities for licences are restricted and a company wishing to proceed with an idea for a new product may have to assign the task to its research and development department. The staff may not, however, possess all the necessary knowledge. Skills and progress will therefore depend upon a company's ability to recruit staff with the necessary disciplines, although this will not be advisable unless there is a certainty that there will be a continuing need for these new skills. The solution is to subcontract part of the problems to laboratories offering a consultative service and should enable a satisfactory solution to be achieved in a fraction of the time required to recruit staff for additional in-house development. Subcontracting may be the only means of foiling a would-be competitor and so avoid the danger of losing a leading market position.

Maddock(9) has commented upon the wide use made of small firms for technical components in the United States where large businesses associate for the specific purpose of establishing a strong base in order to support subcontractors. The latter, unencumbered with a complex management structure, are thus able to pioneer advanced technologies, and the former are free to concentrate on securing markets which are sufficiently large to support the high costs of manufacturing, marketing, and world-wide customer training, distribution of spares, and maintenance. A supporting infrastructure of companies skilled in research and development can also help young, innovative small companies by making knowledge available from their extensive research and product experience. A consequent reduction in product costs, or upgrading of a product specification could help a small supplier, hampered by high costs, to increase turnover and so decrease overhead charges, or could be the means of encouraging a potential supplier to place business in advance of adequate back-up facilities becoming available. If this type of help is not available, a customer frequently provides assistance with cash flow, with the disadvantage that he sooner or later tends to influence the manager, and eventually the supplier will become a dependent company. The symbiosis between the subcontractors and their dominant customers helps to build up a technical infrastructure similar to that sought by the creation of science parks. A somewhat parallel situation in Europe is the continental complex of motorcar manufacturers who rely widely on contract research for engine development and body design.

5.2.5 Mergers

Mergers of companies occur at two levels: the first between a small or moderately-sized company, usually in related businesses, and the second between large conglomerates. The merger of two moderately-sized companies may come about when two parties operating a licence agreement become friendly, and see the advantage of a closer association. In very competitive markets it is used to achieve greater cost effectiveness either by sharing, marketing, and distribution, or other operations.

A merger has two advantages over licensing. It allows entry into each other's markets, both home and export, while patents, trademarks, and other forms of protection are no longer controlled by one party.

The frequency with which large groups come together is high, although the value of takeovers in the United Kingdom fell from £2532 million in 1972 to £291 million in 1975. This source of growth can be extremely successful, and mention may be made of GEC/AEI, Radio Rentals/Thorn/EMI, and Allied Breweries/ Showerings. These mergers are clearly an intrinsic part of the industrial scene but, in common with all business activities, success cannot be guaranteed. A number of investigators have reported on the high rate of disappointing mergers, and Killing(10) has given a figure of 30 per cent downright failures in the United

States during the 1960s. Other articles state that results are usually disappointing, although judgements are too often based on too short a time-scale and allowance cannot be made for the course of events had mergers not taken place.

Since both parties in a merger have comparable powers and rights strategic decisions must, inevitably, give rise to long discussions before consensus is reached. The main difficulty resides in the field of personal relationships. If the two organizations are in related activities, and there is a disparity between the competence of the staff, apprehension will be manifest and there will be considerable pondering upon what the future will bring.

Barlow(11) has discussed the problem in detail, but the basic need is to relieve anxiety by issuing an explicit and comprehensive communication at the earliest opportunity. Staff should be told the reason for the merger, the name of the new company, and important organizational and managerial changes at the same time that this information is released to the press. Thoughts must also be directed towards means of recreating pride and interest. Despite the many difficulties, the number of United States companies of diverse but related businesses has increased steadily since the 1950s.

5.2.6 Conglomerates

An important purpose of conglomerates and aggregates is to mount a defence against environmental risks, and they are built up from diverse and unrelated business. The penalty of remaining in one business sector is illustrated by the Pakamac Company which fell into the hands of the receiver as a result of the exceptionally dry summer in 1983. The only act taken to lessen its dependence upon sales governed by summer showers was to introduce a more fashionable rainwear line(12), whereas it should have diversified into products where sales are stimulated by dry, hot weather. The strategy of hedging is an insurance, not only for single-product businesses, but for multiproduct companies operating in rapidly advancing areas of high technology. The reason for this is that though a company may have a large and excellent research and development department, a chance serendipity by a competitor may result in a catastrophic loss of sales. Noteworthy examples of this may be found among the manufacturers of personal computers. Although diversity inherent in conglomerates is a useful hedge the lack of common strands between businesses can cause difficulties. For this reason a merger, based on synergy, to build up core strength may be a useful prelude to the formation of a conglomerate.

The growth of conglomerates, whose number approximately doubled in the 1970s(13), has been attributed by Channon(14) to the fact that capital intensity is inversely related to profitability. The assets of a takeover can be rearranged and the high price:earning ratio can stimulate other mergers. A conglomerate policy must depend upon the capacity of top management to augment the performance of companies in new fields through delegation, and so achieve an asset:profitability ratio which enhances the stock market image. The higher an industry's average ratio of assets to sales the lower was likely to be the potential return on investment for companies in that industry(14). Firms, therefore, tend to aggregate with the service industries or other activities with a high ratio of sales to net current assets.

The risk of entering areas of new technology, expertise, or cultures, must always be great and a group wishing to build up a conglomerate as a hedge needs unusually high management skill and considerable wisdom.

5.2.7 Joint Ventures

The joint venture, illustrated in Figure 5.3b, is the most straightforward type of diversification through restructuring and arises when two companies recognize advantages which would result from merging one or more of their existing activities to set up a separate unit. The process could also be triggered by a wish to share risks, or because one of the companies, wishing to expand, has insufficient funds for an acquisition, and has been unable to find a suitable purchaser. A more complex situation occurs when a major part of the new operation is substantially new to one or both companies.

Studies by Hlvacek and Thompson(15) suggest that joint ventures most often reflect a high or rapidly changing technology and take place between a technology-based company and one whose strength lies in marketing. The technical company is most likely to be the initiator, even though it may be the smaller. Should one of the two companies be much the larger, the negotiations will normally be found to be longer, more analytical and detailed than is seen to occur between companies of approximately equal size. Joint ventures may be formed by both parties having approximately equal shares or else one party may dominate, as, for example, holding a 70 per cent share. The latter arrangement has the clear advantage that decision making is less likely to be subject to long discussions aimed at consensus. Hlvacek and Thompson also observe that difficulties associated with integrating the activities, especially those connected with personal relationships, are less if the joint venture is some distance away from the originating companies.

Killing(10) has drawn attention to the importance of joint ventures having access not only to current, but to future, technical information. When agreements only cater for exchange of current technology it is usually an indication of a weak relationship between the parties. There is no clear advantage in accepting a minority status in a licensing agreement, and is probably a reflection of a weak bargaining position. Licensing carries a number of risks, of which the most serious is the possibility of the licensor being taken over by another company.

5.2.8 Acquisitions

The reasons for acquisitions are many. A small competitor may have dominated a market niche with a product which a major competitor cannot match, despite his apparently superior R & D facilities, or, alternatively, the acquisition may be a defensive move against a minor threat that has been perceived capable of growth. A manufacturing firm may also wish to diversify vertically by acquiring either a supplier of important purchased components, or a customer who converts the manufacturer's output to products of higher added value. Horizontal diversification may be attractive in enabling a manufacturer to purchase companies in closely related product lines and forms in order to offer his important customers a more complete range of products. Financial considerations may also be dominant, with the object of securing low-growth companies in which valuable assets can be secured at discounted values. This type of acquisition, however, only contributes to growth if the realized assets are put to productive use.

Acquisitions are particularly valuable when the object is to secure a new technology, and Kitching(16) has observed that satisfactory results can frequently be achieved three to five years more quickly than by in-house R & D and design. Interest may also be expressed in companies which are judged to have an undeveloped potential for growth. This can arise when their institutional structure inhibits known entrepreneurial talents of marketing, and research and development, staff. Again, if the owners of small businesses wish to retire they may either have no capable successors or have insufficient funds or talent for the business to continue on its own.

For acquisitions to be successful there needs to be a good fit between the existing and incoming material and human resources, and a willingness of the sellers to impart knowledge and know-how which cannot easily be written up but which are vital to a successful operation of a technology. In common with other diversifications the acquiring company must plan all management moves before the acquisition and attempt to anticipate and counter staff apprehensions. Success is most likely when there is a general recognition of tangible benefits.

A guide to the successful acquisition of unquoted companies and subsidiaries of quoted ones, published by Pearson(17), outlines the commercial strategy and deals with methods of searching for candidates and all matters relevant to their purchase.

5.2.9 Management Buyouts

Management buyouts, spinoffs, or hiveoffs, have become an important consequence of the recession. They may stem from a necessity of large organizations to close a division in order to improve their cash flow, from shareholders in private companies needing to redeem their capital, or from bankruptcies.

In 1983 the estimated number of buyouts was 200, of which some 80 to 90 per cent are thought likely to succeed. This figure is very favourable compared to the 66 per cent normally quoted for start-up businesses. A survey which is to be published by a team for Nottingham University(18) shows a number of encouraging trends: of 103 buyouts surveyed more than 70 experienced an increase in sales since the buyout, more than half exceeded the forecasted profit, and whereas, at the date of the buyout, jobs fell to 83 per cent, a recovery to 95 per cent was achieved in two years or less.

Provided managers have the necessary commercial experience, they can exploit the advantage of close familiarity with the business they wish to acquire, and possess a knowledge of the established customers and market potential. Their greatest asset is a strong incentive to succeed, for their bid for a company will be based upon a conviction that they know how it could have been run more efficiently.

5.2.10 Divestments

A typical divestment situation occurs when the sales of a high-quality product in a large company are lost to a small competitor who quotes low prices through cutting overheads and lowering quality. When such a product is itself a component of a large unit, the customer may have changed the design of associated components, and so permit a downgrading of quality in others. If a market is small and fragmented, it is rarely worthwhile spending capital to regain this type of business.

Divestment is a frequent fate of overseas subsidiaries and can arise from many circumstances. The initial estimate of supporting services may have been grossly underestimated, or unexpected international developments may make it difficult to transmit funds back to the holding company, or political changes may give rise to nationalization and other threats. It also may happen that the availability of raw materials becomes difficult, and importation is prevented by local tariffs.

Divestment may also offer the means of reallocating resources to more fruitful projects. For example, a product might be developed to serve outlets in the engineering business, when unplanned innovation may make it even more suitable for the constructional, or other industries. If the new outlet is likely to become dominant it may not fit in with existing development and marketing skills, and disposal is indicated.

A decision to abandon an activity raises many problems, since the implication is that sales have decreased to a point at which a buyout is impracticable and an external buyer will not be attracted. The activity may be a company within a group, a division of a company or merely a product line, form, or type.

The process of a disinvestment is difficult because staff considerations, sentiment, false hopes, and other subjective feelings weaken financial arguments.

Once employees learn that the future of their company is becoming financially insecure their morale may drop, since they will realize that little new capital will be made available, and prospects for growth are negligible. Once, however, they learn that a divestment is someone else's acquisition they can hope for more favourable prospects, for the situation can be redeemed by a concentrated effort from all employees. Indeed, it is important for employers to convince employees that they may look forward to a more secure future, otherwise key personnel will leave and the value of the enterprise diminishes.

The financial basis for disinvestment decisions has been presented by Choudhury(19) and, expressed simply, a product should cease once its abandonment value exceeds further profits, in other words, once the net present value becomes negative.

5.3 THE RESULTS OF DIVERSIFICATION

The trend towards business diversification in the United States has increased relentlessly over the past 30 years. An analysis, carried out on the 'Fortune' top 500 US companies, shows that from 1970 to 1980 the number of conglomerates has increased from 14 to 34 per cent, and companies with a wide spread of diverse but related businesses increased from 38 to 45 per cent(13). These figures contrast a fall from 40 to 21 per cent in dominant businesses (that is, those in which one activity accounts for 70 to 95 per cent of sales) and from 8 per cent to 5 per cent in single businesses. Of the remaining dominant businesses a large proportion are oil majors. A similar, though less marked, transformation took place in large United Kingdom businesses.

It is difficult to collate literature on diversification because authors do not always distinguish between mergers, conglomerates, and acquisitions. The general conclusion, however, seems discouraging.

Newbould and Luffman(20) looked at the top 511 British quoted trading and manufacturing companies of 1967 and found that of the 59 per cent which survived until 1975 just over half had been taken over. Data on these, collected over the period

1967–72, included shareholders' return on investment, companies' profit on capital employed, and new growth of employees' (excluding managers') wages. It was concluded that companies which did not diversify did better than those which did. The outstanding beneficiaries from diversification were manufacturers whose salaries reflected their enlarged departments.

Studies by Hollier(21) on the outcome of acquisitions showed that expected gains in efficiency were rarely achieved, and other benefits were insubstantial. By contrast Kitching(16) in his study of 27 US multinational corporations who between them acquired 90 European companies, concluded that acquisitions have a great value for companies wishing to introduce or extend their operations in Europe and can produce results three to five years more quickly than by any other route.

Studies which have highlighted the unsatisfactory nature of so many diversifications may not necessarily mean that the principle is wrong, but rather that they have been carried out badly. Furthermore, as was indicated above, judgement must be clouded by the inability to compare what has happened with what might have happened had not the action taken place. The balance of opinion is that because the United States has greater management expertise in the operation of large companies, they have a greater success rate in diversification than has the United Kingdom. Breene and Coley(22) have observed that mergers, for example, are now rarely directed at establishing monopolies and oligopolies, but at businesses outside their core areas for the purpose of regenerating their product portfolios. This change in purpose has given rise to difficulties, and the above authors attribute a lack of success, when it occurs, to one or more of the following four factors:

(1) Acquiring businesses alien to the core activities with extremely few synergistic advantages.
(2) A too-easy acceptance that market growth must mean high profitability(23).
(3) A tendency to pay too high a premium for the acquired company because of a lack of attractive opportunities.
(4) A failure to evaluate in sufficient detail how value will be added to the company.

Past difficulties should not defer considerations of diversification when deciding strategies for growth. The studies should not be regarded as a pointer that failures are inevitable, but rather seen as a warning that diversifications must be carried out by staff of the highest calibre.

5.4 SUMMARY

In the early stages of company growth the required regular increase in sales may be obtained by increasing customer usage, extending existing outlets, and opening up new home and export markets. There will come a time, however, when it will be advisable to develop a portfolio of new products. A too sudden increase in the number and variety of lines, forms, or types, may create management problems, and it is helpful to construct models to see how products relate one to another. This will also serve to demonstrate how current and possible new products fit into the corporate business plan, and may be used to alert the need for organizational changes. A company's need for new products may be triggered-off by a variety of circumstances, and will require every potential source to be reviewed. There are nine possibilities. The first four are inhouse research and development, licensing and technical interchange, and subcontracting, none of which entails the restructuring of a company. The second group of five, by contrast, involves considerable restructuring and comprises mergers, conglomerates, joint ventures, acquisitions, and management buyouts.

When seeking to diversify it will be necessary to consider the extent to which existing skills and experience are complementary. Maps should be drawn based on four synergy types which, together, comprise 15 company functions.

The importance of inhouse research depends upon the level of technology at which a company operates. At high levels it is usually the main source of new products, but at lower levels it has a poor history of success when competition is based on new technology.

Licensing may be a useful adjunct to inhouse research if the former confers the status of preferred supplier in dealings with subsidiaries of large international corporations. An attraction claimed for licensing is that it enables a new product to be marketed without delay, but this is true only when the two businesses are similar. Technical interchange can be a quick means of securing synergistic advantage, but care is needed in translating the economics of a process from one country to another. Differences in culture may also create problems.

If neither licensing nor technical interchange is feasible, subcontracted R & D may be a relatively quick solution and will obviate the need to recruit specialists for which there is no continuing need. It can also be a mechanism by which an infrastructure of large companies can support young, innovative small businesses.

Mergers are a means of obtaining closer association between related moderately-sized businesses to achieve a more effective operation. The main difficulties concern personal relationships, and though the frequency of mergers is high a significant proportion are classed as disappointing.

Conglomerates are an assembly of diverse and unrelated businesses formed to provide a defence against environmental risks. High managerial skills and wisdom are necessary when dealing with a diversity of technical skills, markets and cultures.

Joint ventures are dissimilar from mergers in that only parts of the companies are normally involved. They reflect a high or rapidly changing technology, and frequently take place between a technology-based company and one whose strength lies in marketing. The aim is to complement strengths and eliminate weaknesses. It is important to ensure that joint ventures arrange for access not only to current, but also to future, technical information.

Acquisitions are made for many reasons, and vary from the need to obtain a new technology to the safeguarding of a critical raw material. The fit between both the material and human resources of the buyers' and sellers' businesses should be good, and the latter must be willing to impart information which cannot be easily documented.

Both management buyouts and divestments are a consequence of the recession, and the former is achieving a high success rate. Recent studies have indicated reasons for the discouraging results of diversifications, but it is suggested that past difficulties need not be a pointer to the future.

5.5 REFERENCES

1. Hargreaves, I. (1980). 'The engineer who is putting new sparkle into Coke'. *Financial Times*, **1 October**, p.15.

2. Hargreaves, I. (1980). 'Why Coke is busting out all over'. *Financial Times*, **3 October**, p.17.
3. Pinchot, Gifford (1983). 'Entrepreneurship: How firms can encourage and keep their bright innovators', *International Management*, **38**, No.1, 11–14.
4. Williams, E. (1984). 'The fight for a place in the next video revolution'. *Financial Times*, **7 January**, p.15.
5. Kay, N. B. (1982). *The Evolving Firm. Strategy and Structure in Industrial organization*, Macmillan Press Ltd. London.
6. Ansoff, H. I. (1979). *Corporate Strategy*, Penguin, London.
7. Cooper, A. C. and Schendel, D. (1976). 'Strategic responses to technological threats', *Business Horizons*, **February**, 61–9.
8. George, W. W. (1983). 'Contrasts in management styles - Europe and the USA', *IEE Proceedings*, **130**, Pt.A, No.5, **July**, 288–91.
9. Maddock, I. (1973). 'Sub-contracting - key to technological survival', *New Scientist*, **13 September**, 629–31.
10. Killing, J. P. (1978). 'Diversification through Licensing', *R & D Management*, **8**, 3, 159–63.
11. Barlow, W. (1982). 'The acceleration of change in British industry', *Institution of Mechanical Engineers*, **196**, No.40, 1–6.
12. Hamilton, A. (1983) 'Dry summer seals Pakamac's fate'. *The Times*. **3 December**, p.3.
13. Lorenz, C. (1982). 'Why 'back to basics' may be a short-lived fashion'. *Financial Times*, **24 November**, 16.
14. Channon, D. (1982). *Strategies for the 1980s.* Conference organized by Britain's Society for Long Range Planning.
15. Hlvacek, J. D. and Thompson, V. A. (1976). 'The joint venture approach to technology utilisation', *IEEE Transactions on Engineering Management*. **EM23**, No.1, 35–41.
16. Kitching, J. (1974). 'Winning and losing with European acquisitions', *Harvard Business Review*, **March/April**, 124–36.
17. Pearson, B. (1983). *Successful Acquisition of Unquoted Companies.* Gower Publishing Company, Aldershot, in association with The Institute of Cost and Management Accountants.
18. Brown, M. (1983). 'Buying out becomes big business'. *Sunday Times*, **9 October**, 67.
19. Choudhury, N. (1979). 'The decision to invest', *Accountancy*, **March**, 106–10.
20. Newbould, G. D. and Luffman, G. A. (1978). *Successful Business Policies.* Gower Press.
21. Hollier, D. (1980). 'Companies prepare to net the bargains', *Chief Executive*, **February**, 45–50.
22. Breene, T. and Coley, S. (1983). 'Dilemmas of diversification'. *Management Today*, **September**, 86–9.
23. Wensley, R. (1981). 'The market share myth'. *London Business School Journal*, **Winter**, 3–5.

6

Strategic Planning as a Means of Developing a Business

6.1 GENERAL PRINCIPLES

The contribution of science and technology to those elements of growth which are based upon the maintenance of a portfolio of successful products, is widely acknowledged, and yet it is still not unusual to receive a chairman's report in which product policy is referred to only indirectly, if at all. Fusfeld(1) observed that general business management lacks an intuitive feel for strategically directing and positioning research and development investment, and that one reason for this is the time-span for development failing to mesh satisfactorily with the normal planning period. This must not be taken to imply that innovation policies should be regarded as a separate function and treated in isolation. This would be disastrous. Product development is costly, and decisions on its funding must be taken with reference to every other claimant for resources. Product development interacts with every company activity, and planning for the future must, therefore, involve every company activity.

Business planning became established in the mid-1950s and is now recognized as a discipline in its own right. The proliferation of books and papers on the subject reflects the increasing difficulties faced by businessmen over the last three decades. The process of corporate planning has, however, attracted its advocates and its opponents. The former realized that the astronomic rise in crude oil prices in 1973, and the subsequent recession, destroyed the sense of security which business had earlier taken for granted, and examined ways in which planning could help. They discovered procedures which could shorten board discussions, stimulate new ideas, and enable a more responsive and adaptable corporate attitude to be developed. The opponents, who adopted a pragmatic approach, had their ranks swollen with those who had experienced planning failures caused through sudden and unexpected changes in external factors. It is ironic that failures were not aggravated by planning but by not adopting it soon enough, and then doing it inadequately. Many took refuge in intuitive and supposedly commonsense judgements, and failed to give planning a further trial.

Taylor(2), in a comprehensive review of the literature, regarded corporate planning as comprising two elements, namely, strategic planning and management control. Included in the first are strategies and tactics aimed at a stated company objective. It will include plans for a new product line and a decision on nonroutine capital expenditure. It is this first part that is relevant to this chapter. Strategy has been variously defined, but its description by Gluck(3) as 'an integrated set of actions designed to gain a substantial advantage over competition' has the necessary challenge, while Kallman and Gupta's(4) version is in harmony with much of this argument, namely that 'planning is a continuous process of making present entrepreneurial (risk-taking) decisions systematically and with the best possible knowledge of the future...'. Taylor's wide-ranging paper refers to five different complementary aspects of corporate planning. This chapter deals with two: planning as a central control system for the total enterprise, and planning as a framework for both evolutionary and innovative advances.

This brief review of planning will be restricted to a description of general principles. There is an infinite variety of planning approaches and they arise because of their need to be tailored to the nature and size of a business. The most common period for planning is probably between two and five years, and fits in with the time needed to bring new plant into operation or to develop and launch new products. However, the large international companies often adopt longer time scales. Seidl(5), of the Shell International Petroleum Company Ltd, states that his company adopts a medium time horizon of 10 to 15 years, while Hart(6) of British Telecommunication Research Ltd, states that 'needs research' has shown that advanced technology companies based on capital goods must be prepared to look up to 30 or 40 years hence, although he remarks that this does not exclude a hunch or the flash of genius. Planning should show how strategies and tactics can be formulated to define actions which are needed to attain future goals, and should pre-empt pressures which may be threatened by external events.

Just as the length of a planning cycle varies with the scale of a business and its technical complexity, so does the effort directed to the planning process. An international group is likely to employ a team of skilled specialists who will forecast future trends in technological, economic, and social matters on a worldwide scale and so give guidance on basic principles to their individual companies. The individual companies will then develop their own specific input, relative to their specialized activities, and produce operational plans. Use is often made of external agencies who specialize in producing scenarios of world conditions up to 20 years ahead. These exercises may be extremely elaborate and involve statistical means for analysing and collecting views, on an international basis, of politicians, economists and other specialists in countries all over the world.

In medium to small companies planning has necessarily to be done by executive staff, and is therefore necessarily less sophisticated. During work on the projects referred to above it was encouraging to note that many small companies, managed or owned by staff who had had earlier experience of planning with large groups, had adapted a complex set of procedures to simpler situations with the added advantage of flexibility and immediate relevance.

Unless growth is based upon the best possible data on likely technical advances, emergent market changes, and economic trends, it will carry risk. However, there is general agreement in the literature that planning can help a company to select and adopt the most appropriate actions needed for unexpected changes. This can be done because decisions are translated into financial and operational targets (for example, budgets, and staffing levels). The danger which must be watched for, and obviated, is the one in which planning affects operational decisions in considerable depth and it can assume the inertia of a cumbersome bureaucratic body. A corporate plan must be short, and Pearson(7) has recommended that divisional plans should be restricted to a maximum of 20 pages. A number of planning specialists have enlarged upon other virtues and dangers, of which the more important are:

(1) Planners should be thorough and use sophisticated aids, but yet should retain a healthy scepticism.
(2) Corporate planning should be regarded as a learning system and fullest use should be made of feedback.
(3) Planners should not accept incautiously fixed optimal solutions and ignore creative and intuitive ideas.
(4) Too much reliance on deductive reasoning exposes the company to creative competition.

6.2 STRATEGIES BASED ON THE NATURE OF A COMPANY'S TECHNOLOGY

During a study conducted within a wide variety of manufacturing industries general confirmation was obtained of Nyström's(8) classification (see Chapter 3, p.48), and it was, additionally, observed that a company engaged upon high technology had other features which contrasted sharply with those of a manufacturer of simple engineering components. This situation is shown in Fig. 6.1 in which the vertical axis represents the level of technology, the x-axis the turnover per employee, the y-axis the number employed, and the shaded areas indicate the number of scientists and engineers employed. The high technological company is thus seen to have few employees, of whom nearly half are graduates or postgraduates, and a high turnover per employee. The low technological company has a large workforce whose activities are dictated by craftsmen rather than by engineers, and the turnover per employee is modest. For these two extreme cases the product of the number of employees and their turnover was approximately equal. The construction of these kind of models (see Chapter 3) can help devise strategies for inclusion in corporate plans.

A consideration of Table 3.1 (Chapter 3) and Figure 6.1 leads to the view that it would be worthwhile to explore an intermediate classification and so produce a model of considerable utility. A method of doing this has been described in detail elsewhere(9).

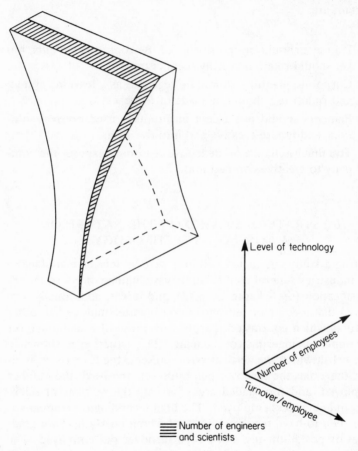

Figure 6.1 A simple model of two business

It is based upon a description of the science and technology which is necessary in order to design, manufacture, and market company products that are competitive. Taking the positional and innovative stances as two extremes, four intermediate levels were defined by linking them with three factors, namely:

(1) The nature of the problem-solving task required to convert an idea into a new product.

(2) The ability and qualifications of research and development staff.
(3) The nature of a publication, if any, which would accept a description of the developments, for example, a trade journal, or a new textbook.

The six levels of technology, which can be identified by all three factors, or, if necessary though with less certainty, from only one or two, were observed to give a general correlation with many company characteristics. Table 6.1 shows those which relate mainly, directly or indirectly, to the process of growth based upon new products.

The above classification is based upon studies in a limited number of companies, but only two misfits were found later when a total of 80 companies were studied. Beginning with companies which operate at levels of technology 1 and 2 it is noted that development is predominantly evolutionary. The distinguishing features between technology levels 1 and 2 are reflected mainly in the quality of staff needed to solve problems that arise during design and development. In instances when the skill and ability of craftsmen were an inadequate basis on which to mount a challenge to competition, the company should make a conscious decision to move up to level 2 and examine other aspects of company characteristics likely to need change.

Companies which operate at levels 3 and 4 practise both evolutionary and innovative approaches. Subgroup 3A contains a higher proportion of the products based on evolution than does 3B, and subgroup 4A still includes a few products based on evolution whereas group 4B does not. Companies at levels 3 and 4 have shown that they have a high aptitude for creating new ideas for products that can be designed by modifying existing ones, and so demand only the minimum contribution from research and development staff. These companies suffer, however, from the disadvantage that their product portfolios are large, with the result that the general production and firefighting problems are so great that new product development has to fight for severely restricted resources. Management needs great organizational and attitudinal flexibility in order to devise strategies capable of supporting efficient existing business without burdening innovation with bureaucratic controls.

Table 6.1 A model for evolutionary and innovative companies

I Level of technology		II New product involves	III R & D costs Capital revenue	IV Risk, benefit	V Sources of ideas	VI Recommended research, development and design attitudes
1		Evolution	Low	Low	Inhouse	Defensive design and development from craftsmen
					and	
2		Evolution	Low	Low	Customer contacts	Defensive design and development from craftsmen and technical staff
3	A	Evolution	Low	Low	Inhouse. Customer contacts	Inhouse R & D from qualified engineer/scientist in conjunction with marketing
	B	A little innovation	Medium	Medium	R & D with marketing	
4	A	A little evolution	Low	Low	Inhouse. Customer contacts	Inhouse R & D from highly experienced engineer/scientist in conjunction with marketing
	B	Innovation	Medium	Medium	R & D with marketing	
5		Innovation	High	High	Inhouse R & D. Contact with customers and places of learning	Aggressive research etc. by engineer/scientist of international repute
6		Innovation	High	High		Aggressive research etc. by engineer/scientist of international repute

VII Necessary company attributes	VIII Recommended paths to growth		IX Main threat	X Employment of scientist/ engineer	XI Number of employees	XII Turnover per employee
High production quantities at minimum costs. Good at distribution and selling	Find new home market segment and new export markets. Improve price and non-price factors relating to product/process. Move up to the next level of technology		Competitors' efficiency at producing and selling. Imports from developing countries	Low	High	Low
				Low	High	Low
Capacity to deal with a wide spectrum of activities. Good at R & D, production, marketing, and selling	Improve sales by attention to non-price factors relating to product	Evolve product	Competitors achieve higher level of technology	Medium	Low	Medium
		Radically improve product		Medium	Low	Medium
Vision and confidence to invest in the future. Excellent at innovating new products and selling	Consolidate world markets and maintain reputation for excellence	Extend product range through radical improvements (Limited market)	Commercial failure by reason of overstretched financial commitment. Failure to maintain world leadership.	High	Low	High
		Market additional unique products outside present range. (Elastic market)		High	Low	High

Groups 5 and 6 are concerned with innovative products that are likely to have been initiated by highly qualified research departments, and certainly need sustaining by these.

Once the level of technology has been identified Table 6.1 suggests likely avenues for growth which can be integrated with their strategies.

6.3 A COMPANY BOARD AND ITS CORPORATE PLAN

Boards should begin their corporate planning by defining the purpose of their business. It should be expressed in a generic form. Haggerty(10) in an account of how corporate planning developed at Texas Instruments, describes how their early efforts to grow caused them to formulate their statement of purpose so as 'to create, make, and market useful products and services to satisfy the needs of customers throughout the world'. This is not as general as it seems on a first, cursory glance, since the three words create, make, and market, are a commitment to innovation.

Another company for which the present author worked for much of his career had, for some 60 years since its beginnings in 1880, seen its role as formulating and manufacturing friction materials. It then realized that this statement was restrictive and changed it into the formulation, manufacture and marketing of (1.) devices which reduced energy to its lowest form and (2.) devices which transmitted energy. By this alteration on emphasis, the boundary for possible new products was extended to embrace mechanical, electrical and other means. In defining purpose it should be recognized to be as an aid to opening the mind to such a wide range of possible innovations that diversification becomes an essential element of company policy rather than an extraneous activity.

The second step of a corporate plan will be to agree goals consistent with the company purpose. To return to the Texas Instrument Company, Haggerty recalls that in 1949, when turnover was $6m a year, they defined their main goal as 'to be a good, big company instead of a good, medium-sized company'. By formulating and carrying out several accompanying strategies they achieved a turnover of $200m a year within 11 years.

The devising of strategies necessary to attain goals requires planners to be well informed on both the human and material resources of their own and competitors' companies. It may be possible to define one strategy of overriding importance. The Texas Instrument Company initiated their progress towards a big company by stating a definite commitment to develop, manufacture and market semiconductor devices initially through three avenues: seeking a patent licence, setting up a project engineering group, and establishing research laboratories heavily orientated towards physics and chemistry of the solid state.

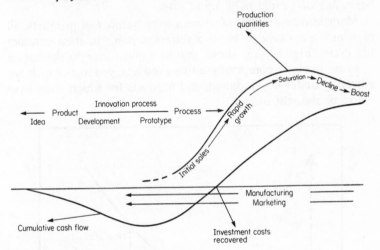

Figure 6.2 Product life cycle

Although goals will vary with the nature of a business, the most successful are likely to be those with which employees can readily identify. General Electric's(11) goal which placed emphasis on competition is in accord with this, and the author's experience, when directing an R & D division, showed that morale could be greatly strengthened by creating a continual awareness of one's own and competitors' successes. Goals are commonly expressed in terms of turnover or profit, when the difference between targets and financial forecasts based upon the extrapolation of past trends is termed the earning or profit gap. If a profit gap is not clearly apparent, it will inevitably become so since all products have a finite life and will decay. The typical life cycle is illustrated in Fig. 6.2(12). Production quantities normally build up rapidly, tail off as customers' needs become largely fulfilled, and then decline either because the market becomes saturated or because the product is challenged by competition, often based on a new technology. The diagram also shows that the innovation process may not be complete at the time of initial sales, and that the automatic decline may be postponed by boosting sales through improved performance associated with lower costs. The cumulative cash-flow diagram indicates the breakeven point when the recovery of the initial investments may be delayed until the beginning of a rapid build-up of sales.

Many companies manufacture a large number of products, all of which are likely to be at a different point in their product life cycle. Fig. 6.3(13) shows that at a given time in the future sales are likely to comprise existing products, products which are currently under development, and products for which ideas have yet to be thought up.

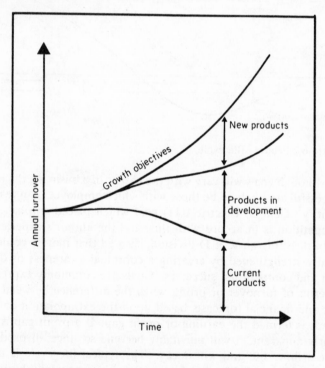

Figure 6.3 Growth objectives through innovation

By indicating the aggregate of individual companies in a number of industries (Fig. 6.4) Wilkinson(14) has shown, in a clear manner, that in the United Kingdom a number of major industries prior to 1930 have declined, industries which came into being after 1930 are approaching their peak and that new industries for the 1980s and 1990s will be required in order to maintain the national output.

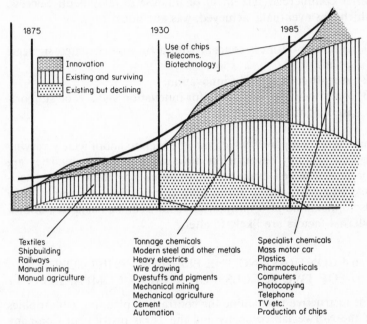

Figure 6.4

Decisions to expand a product range in order to avoid a future shortfall in benefits must thus involve forecasts of a likely lifespan of current products and products under development. Indeed, it is advisable to monitor continuously the status of every product line and to take action when a decline in sales threatens the viability of a particular line. Products whose decline in sales is judged inevitable must be withdrawn, for it is rarely wise to continue after a peak position has been reached.

A further example of a business definition may be taken from a paper on a Norwegian company (Holt) whose goal was a 17 per cent annual growth rate corrected for inflation. The necessary increase in turnover was expressed in terms of current products, products under development and new products, and Fig. 6.3 helped to clarify the strategies. It clearly emphasized the importance of current developments being carried through to success. It also underlined a further strategy whereby the profit gap can be closed by diversifications based on new technologies through either commercial agreement or inhouse development. Success, which was eventually achieved, was attributed to

(1) A sound business concept with a future-orientated strategic policy.
(2) A staff competent in innovation.
(3) An effective organization for innovation and a good supportive information system.

Successful companies which may be seen to adopt widely varying approaches may nevertheless have much in common. They are competent in self-examination, at appraising their present business situation, and in forecasting competitors' strategies. They also understand how far internal factors can be changed and how external factors are likely to alter.

6.4 CONTRASTING THE PLANNING PHILOSOPHIES OF TWO SUCCESSFUL, SIMILAR COMPANIES

It is instructive to examine the contrasting planning philosophies of the Norwegian company and the Texas Instrument Company referred to above.

The Norwegian company's purpose was 'the manufacture in large quantities of technologically advanced components for control, regulation and automation for manufacturers of capital goods'. There were 74 product types, manufactured in seven product lines, totalling 300 items. The company had a product-orientated structure, each with its own research and development, and a central research and development laboratory with a 100 employees. The R & D budget was 3 per cent of turnover. New products were the function of a planning group within central R & D; this was occasionally strengthened by the attendance of marketing experts from outside the company. The group's main concern was to acquire the best possible information on market opportunities, and wide use was made of consultants of international repute. Proposals for new products were based primarily on morphological analyses of existing product lines. The board, to whom the proposals were submitted, decided on whether to acquire suitable licences, to pass to central R & D for the product planning stage, or to refer back for further study. A proposal concerning a new area was dealt with by central R & D. A new product idea relating to an existing product line was passed to the appropriate divisional R & D department whose first task was to confirm that it fitted existing manufacturing, research, and marketing capabilities. If it did, three people were nominated to be responsible for its innovative stages: one for development, one for manufacture, and one for marketing.

The Texas Instrument Company, possibly because its business concerned the highly competitive area of solid state physics and chemistry, adopted a much more complex system. It initiated formal objectives, strategic and tactical systems which defined the strategies and then outlined the accompanying tactics. The system specified a hierarchy of quantitative goals together with a stated performance that had to be measured against a set of specifications. Operational activities, which were clearly separated from strategies, comprised ideas, resource allocation, organization, and reporting. The corporate objective was also supported by nine business objective documents each of which established a strategic organization under a business objectives manager. The scope included appraisal of the market potential, and a projection of technical, market and industry structure trends. It also indicated, in quantitative terms, the parameters against which the organization would be measured, namely: sales, profits, return on assets, and market penetration in terms of percentage of served available market. The business objectives were also supported by strategic statements on opportunity, environment, required innovation, competitive action, contingencies, long-range checkpoints, and probability of success. Each strategy was, in turn, supported by tactical action programmes which described a proposed programme, quantitative goals, resource requirements, responsibilities, and schedules. The company was aware that the formality of planning procedures might inhibit creative thought, and, to make this less likely, they introduced an

'Ideas Programme' for proposed innovations that did not fit into
their objectives, strategies, and tactics system. This programme
had no approved cycles, no delays, no review, and no reports,
and funds were made available up to $25 000 without discussion
or approval. There was additionally a 'Wild Hare Programme'
aimed at identifying and pursuing high risk programmes which,
if successful, would lead to significant and revolutionary types of
innovation.

The Norwegian and Texas Instrument companies have much in
common. Both had a turnover of several hundred million dollars
a year, both allocated between 3 and 5 per cent of turnover to
future activities, and both enjoyed a high growth rate based on
innovation of new products. The degree of complexity in their
planning systems was nevertheless in great contrast. Success is
clearly not linked to a particularly detailed approach. The need
is for an appreciation of the main planning principles, and their
adaption to a particular organization.

6.5 A GROUP BOARD AND ITS CORPORATE PLAN

Even though the holding company of a group may control a number of businesses of quite different character, the strength of their
commitment to a goal is no less than that shown by single, autonomous companies. The difference is that it is more appropriate to have one goal which, expressed in financial terms, can be
readily apportioned among the companies. Argenti(15) has suggested that the one goal, which transcends all others, is 'To make
a satisfactory return to shareholders'. Since the group will normally be owned by its shareholders, who will desire its continuity,
their satisfaction will depend upon the return on capital which
it produces and will produce. Argenti suggests that in a public
company the long-term requirement is for a dividend yield which
steadily grows in real terms. The size of the yield will vary with
the state of the economy but is not difficult to specify at any given
time, although it will vary according to whether the business is an
average United Kingdom one, has international status, or is high
risk. If a group is a private one, the owner's expectations can be
quantified, and will be related to his income or salary, plus his
dividends.

The goal of a steadily growing dividend return implies a number of long-term strategies. Lorenz(16) has contrasted the action of the United Kingdom investor, who rarely thinks long-term, with the German investor who will forego dividend distributions over a period of several years in order that more funds may be allocated to future activities needed to finance growth.

To achieve its financial goal a group must make its expectations clearly known to the subsidiary companies so that they can draw up the necessary goals, strategies, and tactics. Groups will also have duties to its companies, and these, too, should be identified by strategies. Groups may feel it wise to provide central services either because a high degree of specialization is needed, or because by so doing it will achieve a saving in overall costs. Functions which may be administered in this manner include general administration, insurance, legal, taxation, finance, legislation, personnel, management development, marketing, research and development, etc. Should headquarters staff construct possible future scenarios in their planning operations, based on economic, technical and social trends (see p.54), their forecasts should be disseminated to chief executives throughout the group.

A group which fulfils the function of a merchant bank, namely to loan capital to its subsidiaries, must formulate strategies appropriate to the allocation of resources between companies and between short and long-term needs.

It is now widely agreed that group boards have responsibilities both to employees, and those who reside in the vicinity of the subsidiaries. These obligations should be made plain by defining appropriate goals.

Just as considerable variation was noticed in the planning philosphy of single companies, so is the variation in the practice of groups. Allen(11), for example, has described how strategic planning is carried out in the General Electric Company and it is instructive to note that it does not begin with financial targets.

General Electric, aware of an increasing vulnerability to competition, decided that, central to their planning philosphy, there was a need to place emphasis on competition. It developed its challenge under four headings: organizing to focus on competition, developing competitively orientated strategies, emphasizing competition in corporate business reviews and

articulating a clearcut challenge to management. The board identified, as the first step, the design of fundamental changes to the
organization without which planning could not clearly focus on
competition. As a result of an earlier decentralized structure
General Electric had ten groups which were divided, in turn,
into 45 divisions and 175 departmental profit centres; but experience showed that this fragmentation hindered profit centres from
taking an integrated view of their industry. Their answer was to
establish 40 Strategic Business Units (SBUs) all of which had to
meet three criteria:

1. Each was a complete business and included all functions.
2. Every general manager was required to develop a plan which
   balanced long- and short-term objectives.
3. External competition should be identifiable.

and to emphasize the change in emphasis to the competition a
massive training and communication programme was undertaken.

The General Electric annual planning cycle is shown in Fig.
6.5 and involves five separate activities. To ensure that the SBU
plans were competitively orientated the company issued a stan-
dardized competitor analysis drill and a planning practices hand-
book. The first included systematic data gathering and analysis
for seven aspects of competition. The second contained a distilla-
tion of the best strategic analyses and concepts developed during
the year.

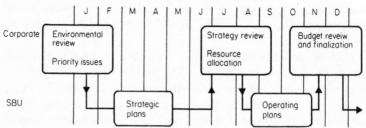

Figure 6.5 The GE annual planning cycle

The corporate body always focused on competition when reviewing plans submitted by SBUs, and Allen remarked that their
main rivals were known to all their employees.

6.6 THE LIKELY EFFECTIVENESS OF A CORPORATE PLAN

In an article on 'Business Lessons from Military Strategy' Widmer(17) lists a number of targets which could perhaps serve as a check on the virility of a corporate plan. Of the questions which may be postulated as a result of Widmer's historical survey, seven seem particularly appropriate:

(1) Would the attainment of the specific goals signify a clearcut victory?
(2) Do the plans aim to capture, obtain and hold markets?
(3) Do the strategies concentrate on place, time, and resources for an attack only after a methodical examination of customers?
(4) In drawing up strategies are assumptions, guesses and future promises discarded in favour of intelligence based on dependable communications?
(5) Are the attacks to be launched with speed upon an unguarded position by infiltration or a flank attack rather than head-on?
(6) Is the mode of attack either based upon surprise through innovation rather than upon the deployment of more resources than competitors?
(7) Is force to be used with the utmost economy?

6.7 SUMMARY

Company growth should be effected through corporate planning, of which two essential elements are strategic planning and management control. The most important outcome will be the building up and maintenance of a portfolio of successful products that will enable the company to sustain an advantage over competitors.

In large companies corporate planning is a discipline in its own right, and may be extremely sophisticated. It is customary to look at least five years ahead, and some companies construct scenarios for a series of lengthening time-scales from two to 40 years. It is important to ensure that the process does not become so bureaucratic that creative and intuitive ideas are discouraged. When corporate planning is practised in medium- to small-sized companies simpler procedures are followed, but the principles remain the same.

A company should precede its planning by defining its purpose and expressing it in a generic form. It should next define its goals. Strategies, of which a number must be long-term, and tactics needed to achieve goals are then formulated and are eventually translated into operational activities. The latter should display the virtues of a well-directed military campaign.

One company goal may be to close the gap that exists between the expected and required profitability, when the appropriate strategies will be concerned with the design and manufacture of new products. Comparison between companies which have much in common shows that although both may adopt identical planning principles they often use means which are in stark contrast but which are equally satisfactory.

A main board should choose goals which can be readily expressed in a form suitable for passing on to subsidiary companies, and, for this reason, financial targets are commonly used. Individual company boards will formulate additional goals, the more important of which will be ones with which staff can readily identify.

6.8 REFERENCES

1. Fusfeld, A. R. (1978). 'How to put technology into corporate planning', *Innovation Technology Review*. Massachusetts Institute of Technology, 51–5.
2. Taylor, B. (1976). 'New dimensions in corporate planning', *Long Range Planning*, **9**, No.6, 80–106.
3. Gluck, F. W. (1980). 'Strategic choice and resource allocation', *McKinsey Quarterly*, **Winter**, 22–33.
4. Kallman, E. A. and Gupta, R. C. (1979). 'Top management commitment to strategic planning: an empirical study', *Managerial Planning*, **May/June**, 34–8.
5. Seidl, R. G. (1979). *How Useful is Corporate Planning Today*, Corporate Finance Conference, October, London.
6. Hart, R. I. (1973). 'Needs research for invention', *The Inventor*, **13**, No.3, 3–11.
7. Pearson, B. (1976). 'A business development approach to planning', *Long Range Planning*, **December**, 54–62.
8. Nyström, H. (1979). *Creativity and Innovation*, John Wiley & Sons, Chichester.

9. Parker, R. C. (1982). *The Management of Innovation*, John Wiley & Sons, Chichester.
10. Haggerty, P. E. (1981). 'The corporation and innovation', *Strategic Management Journal*, **2**. 97–118.
11. Allen, M. G. (1978). 'Strategic planning with a competitive focus', *McKinsey Quarterly*, **Autumn**, 2–13.
12. Piatier, A. (1980). *Les Obstacles à l'Innovation dans les pays de la Communauté Européenne* (Rapports nationaux et rapport général, DG.XIII) Brussels, EEC.
13. Holt, K. (1978). 'Information inputs to new product planning and development', *Research Policy*, 7, No.4, 342–60.
14. Wilkinson, A. (1983). 'Technology—an increasingly dominant factor in corporate strategy', *R & D Management*, **13**, No.4, 245–59.
15. Argenti, J. (1975). 'Setting objectives. A practical approach', Chapter I in *Corporate Planning and the Role of the Management Accountant*, Institute of Cost and Management Accountants.
16. Lorenz, C. (1979). *Investing in Success: How to Profit from Design and Innovation*, Anglo-German Foundation. London.
17. Widmer, H. (1980). 'Business lessons from military strategy', *McKinsey Quarterly*, **Spring**, 59–67.

7

Finding ideas for improved
and new products

7.1 THE SEARCH FOR IMPROVED, NEW,
AND UNIQUE PRODUCTS

The extent to which a company may obtain a substantial advantage over competitors is a measure both of the strategies and the manner in which they are accomplished. The owners, or board of a company, will need to decide which markets merit withdrawal, continuing unchanged, strengthening, a major expansion, or entering for the first time. Of these strategies a number will involve development within the company of modified and new products and their success will depend, crucially, upon the worth of the

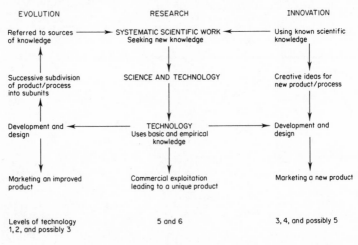

Figure 7.1 The processes of evolution, research, and innovation

Table 7.1   Possible sources of ideas

I.   Desk research

    1.   Relevant trade and technical journals and abstracts
    2.   Patent lists
    3.   Competitors' and suppliers' catalogues
    4.   Newspapers, such as *Financial Times* technical page
    5.   Scientific journals
    6.   Sources of statistics and economic intelligence
    7.   Directories, such as *Kompass, Kemps,* US and Canadian Marketing Surveys
    8.   Monitoring new components, particularly new materials, new processing methods, new chemicals

II.   Inhouse sources

    1.   Company suggestion scheme
    2.   Staff in market, research, development, and indeed all departments
    3.   Review of shelved projects
    4.   Evolutionary development
    5.   Innovatory development
    6.   Basic research

III.   External contacts

    1.   Present and potential customers and suppliers, competitors
    2.   Trade associations, research associations and professional institutions
    3.   Universities, polytechnics, and consultants
    4.   Licensing consultants, patent brokers, and data information banks. University industrial liaison services
    5.   Technology exchange fairs, exhibitions, conferences. 'Open days' of relevant establishments
    6.   Department of Trade and Industry's information and technical services

ideas behind them. The aim must always be to market products that are virtually unsurpassable. There are so many sources of ideas, of which a selection is shown in Table 7.1, that it is difficult to decide how best to carry out a search. This is a topic which has not been adequately dealt with in the voluminous literature on innovation, but as a need clearly exists, an approach will now

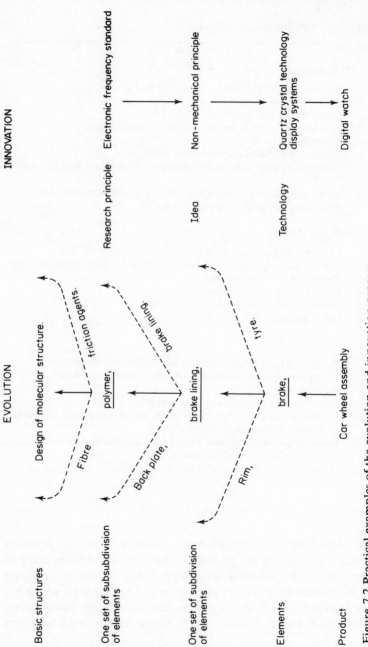

Figure 7.2 Practical examples of the evolution and innovation process

Table 7.2 Ways of achieving successful products

Characteristic features	The route to new products		
	Evolution	*Innovation (general)*	*Innovation (research)*
Goal	Improved products	New products	Unique products
Main focus of ideas	Manufacturing units reacting to customer and market needs	Marketing and sales in association with R & D	R & D
Main activity	Analysis of own and competitor's products, and application of appropriate technology	Creative ideas for new product/processes, and application of known scientific and technical knowledge	Discovery and application of new scientific and technical principles
Means of Stimulating Ideas	Workshop designed to explore systematic methodologies	Workshops designed to stimulate creative ideas	Contact with external centres of excellence (constraints too restrictive for formal workshops)
Levels of technology	1, 2, partly 3 and 4	Partly 3, and 4	5 and 6

be described whereby a company may direct its attention to those areas which are likely to be the most fruitful.

Figure 7.1 illustrates the contrasting features between three routes to new products, namely evolution, basic research, and innovation, between which there will always be considerable interaction. The evolution method is illustrated in the left- hand column. Improvements are first sought by analysing the function of a product's constituent parts, and followed by a consideration of those elements of which they are composed, a process of subdivision which is continued as long as it produces useful results. The number of times this process is carried out depends upon the initial complexity of the product, and may be taken to the point at which consideration needs to be given to molecular structures. With each successive subdivision it becomes increasingly likely that science rather than empirical development is involved.

Innovation, shown on the right-hand side, draws its inspiration from existing knowledge. The initiating concept may be stimulated by commercial need, but the important stage is one which concerns both development and design. In Fig. 7.2 are two practical examples: the first represents a programme designed to improve a car wheel assembly, and the second indicates the likely principles involved in the invention of a digital watch.

Table 7.2 advances the analysis, and shows that in the continuum from evolution through innovation to research the focus of ideas moves from manufacturing through market to the laboratory. The main activity similarly passes from short-term, empirically based, technology through creative medium-term research and development, to long-term original scientific studies. The levels of technology associated with each route are shown.

The above three routes and their characteristic features form a basis on which to carry out a search for new ideas. The more important sources will be discussed and followed by a description of consultancy sessions (termed workshops) that were conducted in a number of companies for the purpose of stimulating evolutionary and innovatory ideas. (see Introduction, p.xvii.) This particular approach is considered only to be applicable for levels of technology 1 and 2 and possibly 3 and 4. Levels 5 and 6, which involve research of a highly specialized nature, are dealt with in Chapter 8.

7.2 EVOLUTION

7.2.1 Customers and suppliers

During the early days of X-ray crystallography, infrared spectroscopy, and many other analytical techniques, prototype instruments were designed and made in university laboratories, and several years usually elapsed before demand was sufficient to attract commercial interest. This practice of designing one's own instruments spread to industrial users who made a practice of submitting detailed specifications for improved instruments to manufacturers. Studies by Utterback(1) and Hippel(2) have confirmed this trend, not only in the instrument but also in the

	Manufacturer's knowledge of Customer's need	
	Poor	Good
Customer's need		
Current	Customer Initiates Action	Customer and/or manufacturer takes initiative
Latent	No progress	Manufacturer takes initiative

Figure 7.3 Likely source of ideas — industrial customers

process industry. Hippel's work showed that the percentage of user ideas adopted by manufacturers of instruments and process plants was 80 and 60 per cent respectively. He regards the contribution to innovation by users as a feature which distinguishes companies supplying the consumer market from those which sell to the industrial companies. Figure 7.3 is reproduced, with minor amendments, from a paper by Tushman and Moore(3) and shows the importance of a manufacturer being familiar with a customer's current and latent needs. It is interesting to note that Parkinson(4) has attributed the poor performance of certain British industries, compared to their German counter-parts, to our reluctance to become involved with suppliers. It must, however, be remembered that unless manufacturers' responsiveness is matched by considerable design ingenuity, the penalty of a too extensive product range will be incurred.

Wearden(5) has suggested that a search should not be restricted to customers but based upon a network of customers, competitors and suppliers, as shown in Fig. 7.4. Where possible an analysis

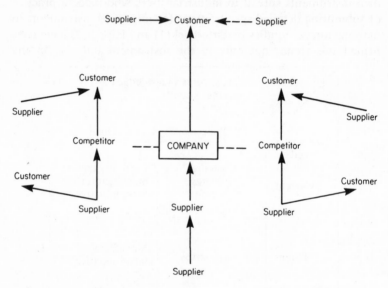

Figure 7.4 A product search based upon customers, competitors and suppliers

of the product range should be compiled for all of the 15 or more companies. Wearden suggests that where competitors have diversified in a rational manner, an inspection of their catalogues should stimulate ideas for compatible products, while inspection of suppliers' products should, additionally, reveal opportunities for vertical integration.

7.2.2 Product analyses

If a product has been marketed for several years, it is wise to consider the extent to which it still meets its intended purpose.

The original product specification may need to be tightened in some respects and relaxed in others. If, for example, the life of a product exceeds practical requirements, but the cost is too high, the answer may be a simpler design. It is also worthwhile carrying out a field study of product behaviour under operating conditions when any anomalous behaviour can be observed. Acute observation is essential since such behaviour is apparently illogical and tends to be seen but not heeded. Anomalies, however, have led to many world-famous discoveries, and once they can be explained may yield original ideas of considerable potential.

If value engineering has not been used in the design of a product, a value analysis type exercise is recommended. A small group of carefully selected people, under a trained leader, answers four questions: what is it? what does it do? what does it cost? and how and when can it be done better? Where high costs are shown to be an important factor, the group should also conduct a functional cost analysis exercise(6). Following an exhaustive examination of a product and its behaviour, thought should be given to the parts of which a product is comprised. The contribution which each makes to the performance of the product should be known, and the more important could be subjected to a value analysis exercise. Once the function of each part is fully understood, a morphological matrix (see p.162) should be constructed for the purpose of finding better alternatives for all, or a proportion of, the elements. New methods of assembling parts may also repay attention since the original development work may have introduced constraints which have restricted performance or affected costs.

In most industrial products, mutual interaction occurs between many of the constituent parts, and to change one at a time is likely to give misleading results. Statistically planned experiments must be used, and a number of changes should be introduced simultaneously in order to uncover interactions(7).

7.3 INNOVATION

We have seen (p.46) that innovation differs from evolution in the magnitude of the advance, and its consequent effect on the company. The sources of innovation can be so varied that it is proposed here to refer to them under the headings of 'general' and 'research', shown in Table 7.2.

7.3.1 General

Companies operating at levels of technology 3 and 4 differ in respect to the emphasis they give to evolution and innovation. These companies form the majority in the United Kingdom, and are characterized by large product portfolios, a bias towards batch production, and a high degree of complexity. In the more successful companies marketing or sales closely collaborate with development staff, and there is a tendency to pursue an aggressive product policy(8). Under these conditions there are two main sources of innovative ideas. Either marketing staff recognize new business opportunities from discussions with their external contacts, or development staff proceed so far down the path of element subdivision that they either subcontract or initiate inhouse research that leads to new concepts. Although these sources should be augmented by the normal development function, this does not happen as often as might be expected. The large product range precludes a high degree of specialities, and should a company enjoy a dominant market position it rarely explores venturesome ideas(9).

7.3.2 Research

At levels of technology 5 and 6 research and development are the main source of innovative ideas, and should be capable of ensuring company growth. Unfortunately, highly specialized work is too often likened to evolution, and it is salutary to recall Dörfel's review(10) of innovation in the chemical industry, when he remarked:

> Unlike most of our other industries, the chemical industry did not primarily arise out of crafts and trades, but as a result of constant translation of scientific research data into commercial-scale practice. New developments in the chemical industry were characterized by a scientific, systematic approach and were much less the result, as was the case

in many other industries, of mere empirical experimentation. Once the scientific theoretical principles underlying the laws of chemistry had been worked out, new developments were made more quickly and often represented a big leap forward.

The late 1970s and early 1980s have seen a general denigration of the contribution of research to new products, and until this trend is reversed rich potential rewards will continue to be passed by.

7.3.3 The use of creative groups to generate new ideas

In Chapter 2 reference was made both to group behaviour and to hypotheses on creative thinking. Groups set up to stimulate creative thinking have been used for a long time, and the most widely used method, synectics, was described by Gordon(11) in 1961, and Parker(12) has commented favourably upon the application of this technique within an industrial context.

A brief description of the process written by Vincent Nolan, of Abraxas Management Research, is given in Appendix I, p.172. Holt(13) has long advocated training in creative thinking, and regards idea generation as a key to successful management of change. Many different approaches have been developed, particularly in Japan and Germany; and Geschka(14) of the Battelle Institute, has classified some 50 formal heuristic principles into four groups. These are briefly described in Appendix II, p.174.

Increasing use of idea generation groups is being made by companies who operate at low levels of technology and manufacture consumer products. At intermediate levels of technology, creative methods have also been successful both in uncovering new business opportunities and in solving apparently intractable problems. They are rarely used at the highest levels of technology.

Personal experience suggests that creative methodologies can be effective in all but rare instances, provided three obstacles are overcome. The first stems from the rational, analytical discipline of the scientific method, which results in practitioners becoming insensitive to the importance of inspiration. Scientists tend to be insular, introvert, and believe that to seek help is an admission of failure. The exceptionally creative scientist can rely on his prowess and only requires informal contact with his peers, but a research director cannot rely on appointing genius and must seek to raise the level of the collective intellect of his staff. An important tool is the use of formal groups trained in problem solving and in recognition of new opportunities.

A second obstacle is the time needed to acquire skills, and Parker observed that after two weeks training, and six months' experience of operating weekly half-day synectic sessions on industrial problems, the performance of the group continued to improve.

The third barrier lies in the selection of group members. At levels of technology 1 to 3, and possibly 4, the need is to represent a wide range of interests ranging from laymen to one or two specialists. At levels 5 and 6 all group members must be capable of understanding problems of high complexity and should represent a wide range of specialization.

As a consequence of these difficulties highly qualified scientists in particular are prone to decry group techniques, and are unlikely to be convinced otherwise unless they gain first hand experience by taking part in a session led by an experienced and competent practitioner.

7.4 OPERATION OF WORKSHOPS DESIGNED TO GENERATE IDEAS

One phase of the investigation described in the Introduction was concerned with helping smaller companies to increase their product range, and it was found that success depended upon teams having considerable knowledge about both the firm's and competitors' products and the commercial situation.

Experience showed that it was advisable to conduct two workshops. The purpose of the first was to take stock of all material and human resources. The latter often disclosed unsuspected talent, and in companies which had not recorded know-how and knowledge exercised by staff in previous situations, discussions on the subject were held with key members of staff. The second workshop was built upon the first, and used creative groups to discover how ideas for the future could flow from knowledge of the present.

7.4.1 Workshop A—the present

The workshop should consist of a morning and an afternoon session and be completely free from outside interruption. A convenient arrangement is to use a local hotel and arrange for a working lunch. To create an informal atmosphere easy chairs should be arranged in a semicircle facing the leader, and it is important that there should be no tables. Every idea is entered on a flipchart which is then displayed on the wall for future reference.

The group should not exceed eight in number, and in addition to the leader, should also comprise the chief executive, divisional heads responsible for marketing, and/or sales, development, manufacturing, and finance. Eight procedures are detailed below, and all serve a specific purpose.

Procedure no.1

Table 7.3   Some aspects of the current business situation

Where are we now?

Profits	Just breaking even
Cash flow	Adverse (cost of tooling/stocks!)
Products	No uniqueness
Brand image	Good, except for product 'A'
Market outlets	Home: 95 per cent Export: 5 per cent
Competition	Fierce
Prices	We are leaders; added value variable
Service	Very good
Technology	Equal or better than competitors, but not achieving its potential
Production capacities	More than sufficient
Production methods	Fairly good—but a lot of hand assembly
Selling methods	Wholesaler 50 per cent
	Retailer 25 per cent
	Manufacturer 25 per cent
Resources	People/skills—all right

In this exercise the group is asked for a general indication of the state of their business. Only a few possible headings should be suggested, for it is necessary to discover which subjects are considered to be important. Individual contributions will vary, but consensus is sought, and when obtained the outcome is written up on the flipchart. Table 7.3 shows a typical example and this, and subsequent illustrations, are taken from one workshop.

Procedure no.2

Table 7.4   Innovation audit

---

1. Is the company exploiting market segments in which they are currently engaged?

2. Are there opportunities to enter new home and export segments?

3. Are there possibilities of expanding the market by relatively minor product improvements?

4. Which of the company's products are unlikely to show increasing sales?

5. When will new products be required and what contributions will they be required to make to future turnover and profits?

6. Is the company in a position to allocate sufficient resources to new product development?

7. What methods are used by the company to seek ideas for new products?

8. How does the company make a final selection of new product ideas for development?

9. Who will be responsible for planning and monitoring progress?

Exercise No.1 rarely discloses a company's attitude to product innovation, and the group is asked to discuss and complete a questionnaire, of which an example is shown in Table 7.4. Experience has shown that by interposing specific, between more general, routine questions, group interest and energy can be maintained at a high level.

Procedure no.3

This exercise is used for two reasons: to gauge the group's sympathy with long-term planning, and to introduce those members, who are to attend the second workshops, in the use of visual imagery. Members are asked to select a site within their company, and visualize how it would appear on the assumption that they had achieved all their ambitions. The time should be five or ten years in the future. Irrespective of whether a small room or a large factory had been chosen, the imagery should be constructed in detail and be seen as a vivid coloured miniature against a dark background. The scene must be one that is desired rather than one which is believed to be achievable.

Each member of the group is invited to describe his scene in great detail, and include reference to all the senses. If the imagery is seen to be inadequate the exercise is repeated. Individual contributions are consolidated into one picture which harmonizes the individual aspirations and are thus translated into new company requirements. An illustration taken from a workshop is shown in Figure 7.5 and can be compared with Table 7.3.

Money needed for new products for the consumer market.
The company to be identified with product uniqueness.
Four or five generic groups to be found with high growth potential.
Develop a product which . . . .
Find consumer products which are not electronic involving information/
    communication/automation/computer.
Marketing – aim to be a world leader for a wider audience.
Include products with a short life and high success rate.

NEW REQUIREMENTS

MORE
RESOURCES
FOR
    / Innovation
    — Market research
    \ High level automation

Use of computer databanks
Direct mail facility
More export business
Bring factory and warehouse together on one site.

Figure 7.5 Wishes for the future

In a final stage the group advances ideas whereby the dark background can be crossed to attain the central image. These are then converted into strengths and weaknesses and used in procedure no.6.

Procedure no.4

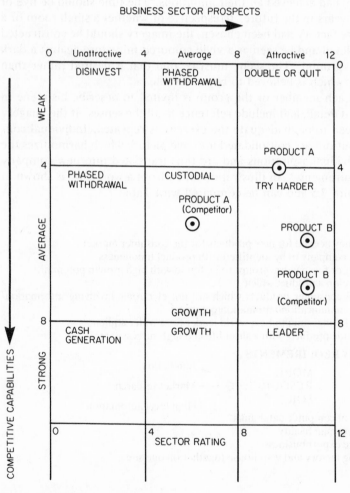

Figure 7.6 Directional policy matrix

A directional policy matrix is constructed to help groups compare their own and competitors' products, and is a crucial stage of the general process of idea generation. The form described by Robinson, Hichens and Wade(15) is recommended and requires a group to agree, for each product under consideration, a numerical value for its business sector prospects, and competitiveness. Figure 7.6 shows a completed matrix in which the diameter of the circles is proportional to the turnover of the product. A group may choose to include considerations of market growth, market quality, and environment in the business sector prospect and market share, production, and development in a company's competitive capability. Market quality and production each comprises three elements: profit record, stability, past history and capacity, plant, and labour respectively.

The description of each box within the matrix refers primarily to strategies, all of which have some relation to product development. In the example given in Fig. 7.6, product A has a business data rating of 6 and a competitive capability of $5\frac{1}{2}$. Its position in the 'custodial' box indicates that effort should be directed towards maximizing cash generation for the purpose of developing a unique product with the aim of becoming a market leader. The 'try harder' box suggests that the product is in danger of becoming very vulnerable. With a competitor in the same box, a struggle for supremacy is clearly indicated.

In this particular example, when the competitor shares a similar situation, these procedures present an ideal opportunity for reviewing strategies in an unconstrained, receptive atmosphere. By confining the discussion to options open to the competitor, staff will not be reluctant to offer suggestions because of their possible adverse effect on interpersonal relationships. The final analysis need only be tailored to fit the company situation during the closing stages of the session.

Procedure no.5

In Fig. 7.6 both competitors' products show the higher competitive capability, and when this happens the respective merits of products are highlighted by comparing product attributes

Table 7.5   Typical list of attributes

Technical performance
Tightness of specification
Fitness for purpose
Aesthetic quality of design
Functional quality of design
Ease of use
Durability
Reliability
Cost level
Compatibility
Low cost replacement parts
Ease of installation
Ease of maintenance
Ease of storage
Delivery times
Service
Ancillary functions
Distribution

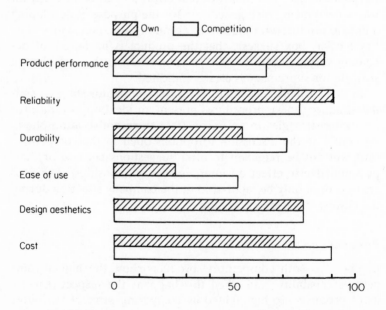

Figure 7.7 Attribute listing

Groups are asked to discuss each attribute and position it on a nominal scale which ranges from unacceptable to perfection. Table 7.5 shows a representative list of attributes, and Fig. 7.7, due to Fusfeld(16) shows a convenient visual display.

This procedure quickly brings under review most aspects of a business, since justification for, and objection to, proposed ratings, will be substantiated by detailed information. To conclude this procedure every competitor attribute which achieved a high rating is arranged in descending order of magnitude, and the group is asked to consider in relation to their own products: (1.) which attributes are likely to be improved by development currently in hand; (2.) what new work could be initiated to deal with the remaining attributes; and (3.) which attributes are known to be the subject of work by competitors.

Procedure no.6

Table 7.6   Where will we be in five years' time?

1. Turnover and profit
2. Number of unique products
3. Market outlets (home and export), and customer relations
4. Market share of principal products
5. New technologies and management skills
6. Dependence on suppliers, licensees, and customers
7. Production capacity, delivery times, material utilization, utilization of plant, and batch size
8. Distribution networks
9. Material and human resources
10. Cost of labour

At this stage a group should have a sharpened perception of the present situation, of their hopes for the future, and of the shortcomings of their products. It remains to compile a list of special skills that exist in the company, together with a brief statement on both the important human and material resources.

The group is then asked to postulate a more formal description of a five-year goal. Headings should differ from those used in procedure No.2 (Table 7.6) and should reflect experience gained in the preceding exercises. Members are again asked to suggest additional headings and encouraged to be original, constructive, and to desist from making value judgements.

Proposals are finally agreed by the group and are then converted into a formal list of strengths and weaknesses. Most companies will have already incorporated a list in their corporate plan, with which a comparison can be made. If the workshop has been successful there will be little in common between the two statements, and company activities will be regarded in a fresh and novel manner.

Procedure no.7

Table 7.7 Functional and morphological analysis

I Component part	II Own product	III Competitor product	IV Alternative suggestions
Body			
(a) Manufacture	Die pressed	Die cast	Forge, turn,
(b) Material	Steel	Aluminium alloy	fabricate plastic, copper, cast iron
(c) Design			
(i) Cover	Hinged cap	Hinged cap	Plug, shutter, press-on cap
(ii) Flange	Circular 15 cm diameter 5 mm thick	Circular 140 mm diameter 3 mm thick	Reduce diameter increase diameter hexagon shape
Sensor	Electro magnetic	Inductive	Capacitous, fibre optics, mechanical

Where competitive products exhibit a number of superior attributes a search should be made for possible evolutionary modifications. At levels of technology 1, 2, and 3 success frequently follows an application of morphological analysis(17) which may then lead unexpectedly to major innovations.

An analysis used in one workshop is shown in Table 7.7. Column I shows component parts of the products, column II the nature of the company's part, and column III the parts used by the competitor. The last column lists the alternative suggestions that were obtained in a brainstorming session. If a selection is made of the best idea for each component part, and all feasible combinations are examined, it is usually possible to suggest ways of achieving the required improvement to an attribute.

Procedure no.8

This is the last stage of the workshop and each individual is asked to give his assessment of any new idea which he believes could enhance the company's competitiveness and, in particular, with respect to product development.

Each member of the group is then given a table (Table 7.8) which suggests a sociological trend, enhanced or diminished opportunities, and possible, relevant industrial activities. The group should be invited to discuss this table, add to it, and select both technologies and markets relative to their current, and possible future, activities.

At this point those members of the group who will be participating in the second workshop will have gained the necessary background knowledge without which their performance is unlikely to be outstanding.

7.4.2 Workshop B—the future

The purpose of this workshop is to reinforce existing entrepreneurial qualities of creative staff by combining creative thinking with a logical approach. The intention is to find a few ideas whose merit is so great that idea selection is easy, and production and marketing straightforward.

The workshop should comprise nine members of staff, in addition to a trained leader, in order that it can be subdivided, when required, into three teams of three. It should include two from the previous workshop, and women should be included

Table 7.8 Future Activities

Group	Sociological Trends	Enhanced/(Diminished) Opportunities	Relevant Activities
1	Growth of world population	New energy sources	Oceanography and offshore development. Nuclear energy. Mineral extraction.
		Conserving and storing energy Alternative materials	Recycling and reclaiming processes. Thermodynamics. Development of both new alloys and plastics. Fibre reinforcement.
		New food sources	Agriculture. Microbiological insecticides. Genetic engineering
2	Ageing population	Preventive medicine. Home care. Transplants Leisure Shrinking teenager market	Pharmaceuticals. Biotechnological diagnostic methods. Surveillance systems. Surgery. New transport systems. Reduction of 'trendy' disposable consumer goods.
3	Shorter working week	Leisure industry Education facilities	Sports gear and clothing. DIY equipment. Horticulture, Photography. Transport systems. Home computers. Telecommunications. Knowledge engineering.
4	Increasing indiscipline	Protection of people Protection of property	Elimination of cash transactions. Alarm and surveillance systems. Fire detection and prevention. Anti-riot equipment.

5	*Elimination of repetitive and dirty jobs*	Automation	Automated offices. Automated sorting, conveying, and disposal systems. Multiprogramme robotics assembly. Automation of manufacturing and extraction processes. Microelectronics.
6	*Higher quality of life*	Aids for study and learning	Telecommunications. Personal computer databanks. Video displays. Word processors. Printing and publishing. Elimination of pollution.
		Health	Heating, ventilation, and air conditioning. Biotechnological aids for maintaining health. Provision for physical exercise
7	*Greater inter-action between societies*	Simpler and cheaper communications	Telecommunications. Microelectronics. Fibre optics. Information technology. Automatic financial transactions.
8	*Increased production from less developed countries*	Export of sophisticated processing plant	Flexible manufacturing systems. Computer-aided design and manufacture.
		Export of high value-added products	Research, development, and design. Automated, analytical laboratory equipment.
		Higher exports of low weight high cost products	Research, development and design at higher levels of technology.

because they are capable of imaginative insight of a different nature from that normally shown by men. The participants should be outgoing and creative, and success is more likely should they have hobbies at which they are outstanding. A typical team would include the managing director, the marketing or development director, a senior and a junior manager from both marketing and sales development, a staff member of the works department, a member from the shop floor, and a representative from the secretarial or typing pool. It is important to have at least four members who are not constrained by having a too detailed knowledge of the company's products.

Informality is again important if ideas are to flow freely, and a friendly atmosphere must be established in which status is not an inhibiting factor. A leader should develop his own introductory routines for achieving this, and he may also find it necessary to introduce the group to brainstorming and brainwriting exercises.

Although only two procedures are described below, a large number of alternative creative methodologies are available (see page 154) but most require leaders who have had training in particular skills.

Procedure no.1

This can be one of the creative group exercises referred to in Appendix I, of which the best known is synectics. It is particularly appropriate for companies supplying the consumer market and for those companies operating at technical levels 1 and 2. There is the obstacle that a leader with the appropriate skills and training is required, but once overcome it should result in exciting suggestions for new products.

Procedure no.2

This procedure is recommended for companies operating at technical levels 3 and 4, and requires the construction of a 3D model of the company's business, published by Carson(18). The form of the model is shown in Fig. 7.8, and consists essentially of one axis which represents a market segment, or industrial sector, and two axes which characterize existing products. The original suggestion for the latter two parameters were 'raw material', and manufacturing process, but this was found to be restrictive and the concept was later broadened by Carson and Rickards(19) when the axes were extended to deal with other than existing products, processes, and raw materials. The choice of parameter is crucial to success and demands considerable thought. In one workshop the product was successfully characterized by one axis showing 'initiating input', for example, liquid level, light, sound, infrared, touch, etc. and the other listing 'controlled output', for example, linear motion, rotary motion, high frequency pulses, etc. A more complex problem occurred in a company whose production experience was limited to a non-corrosive alloy and one process. A list was made of design skills, and a synectics session provided a generic group of possible engineering components for one axis—precisely positioned, lightweight, strong structures and complex corrosion resistant underwater structures; and a list of possible, generalized, applications for the second axis—all types of filtration and separation systems, and reinforcing structures for non-metallics.

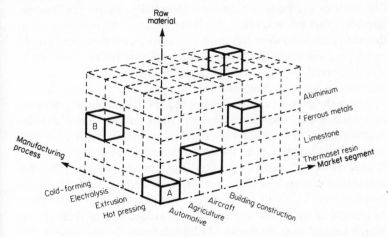

Figure 7.8 A model to highlight possible marketing opportunities

Immediately a decision is reached on the axis parameters, the team is subdivided into units of three, and each given a 2D matrix corresponding to one market segment and asked to suggest products for each box. Once the generation of ideas begins to slow down, the teams are given a second sheet for a different market.

This method should yield as many as a hundred ideas for two market segments within four hours, and a company can then run further sessions until 500 to a 1000 ideas are collected. A few ideas should win widespread support, and each group member should be asked to select the most promising six. The final selection must be made by the executive responsible for new products, who should always be included in the workshop. Products should be capable of being developed with existing resources if the axes have been selected with this aim. When it is felt that a company should acquire additional resources, the axes can be suitably extended. Carson, who aptly described the process as 'finding the needle in a haystack by searching through every bale', could have added that the haystack is an expandable one.

7.5 DESIGN METHODOLOGIES

The result of introducing well-known design methodologies into postgraduate project work, as part of the design course at Loughborough University, has been reported by Pugh and Smith(20). Qualitative methods which are essentially simple in nature, namely, analogy, inversion, attribute listing, and non-numerical decision matrices, were found of value provided they were taught by those competent in their use, but quantitative creative methods conferred an unwarranted degree of certainty that tended to discourage students from further thought.

From the above experiences Pugh(21) has evolved a systematic approach which, despite its logical basis, nevertheless includes many recognizable elements common to many qualitative creative methodologies. The method is exceptionally thorough, and is analagous to Carson's matrix of business opportunities in that a systematic search is made for the outstanding idea. In an example given by Pugh, consultancy design problems were given to groups of up to seven postgraduate students over their course which ranged over 46 weeks. Although tutorials and other activities were included in the course, the time allocated to the design problem was considerable.

Should this methodology be used in conjunction with other activities, care must be taken to direct sufficient effort so that the optimum solution can be achieved rather than merely a satisfactory one.

A project should begin with a decision on the boundaries of the design, and groups should familiarize themselves with all likely aspects, however remote. An exhaustive survey of the literature should be undertaken when a group may, typically, need to peruse many hundreds of articles and several books. Should the requirement call for a new concept of an existing product, every competitor's product should be obtained, examined, tested, and analysed. Armed with the knowledge obtained by the above two procedures, a questionnaire is drawn up for discussion with the client, and so enables a detailed, explicit specification to be agreed.

A parametric analysis is then carried out and is concerned primarily with finding out whether or not relationships exist between the many parameters for the product under consideration.

On page 56 an example was given of a parametric relationship, namely, the peak energy dissipation of a brake lining plotted against time. A considerable number of other crossplots were explored during the studies, many of which proved to be of practical value. The number of brake applications plotted against speed was related to the power of the engine and the terrain, while use was made of plotting the number of brake applications, as a percentage of the total, against five variables measured at the brake discs while the brakes were applied: energy dissipated, maximum rate of working, mean rate of working, maximum deceleration and mean deceleration.

Pugh states that hundreds of parametric plots may be necessary. The number which may prove useful will vary, but typically, out of 100 to 200 plots as many as 20 may be relevant to a design study. This type of exercise will normally enable many relationships to be converted from a qualitative into a quantitative form.

The next phase involves the creation of likely concepts based on the general specification and the derived quantitative data, and as each one is generated a large sketch is drawn to indicate the essential principles; this is displayed on the wall of the design office in order to encourage the interaction of ideas. This part of the design process may be spread over several weeks when some 30 to 40 different ideas may emerge. To decide which concept is the best is regarded by Pugh(22) as

CONCEPT / CRITERIA	1	2	3	4	5	6	7	8	9	10	11
A	+	–	+	–	+	–	D	–	+	+	+
B	+	S	+	S	–	–	A	+	–	+	–
C	–	+	–	–	S	S	T	+	S	–	–
D	–	+	+	–	S	+	U	S	–	–	S
E	+	–	+	–	S	+	M	S	+	+	+
F	–	–	S	+	+	–		+	–	+	S

Figure 7.9 A concept comparison and evaluation matrix

the most difficult, sensitive, and critical of all design problems. The chosen concept must not only be the best possible within the constraints, but must be thoroughly understood so that alternative options can be refuted by sound technical argument and debate. The recommended method is to construct an evaluation matrix in which generated concepts are compared, one with another, and against criteria which, based upon requirements detailed in the product specified, are chosen for evaluation. One concept is chosen as the datum, which may refer to a competitor's product, and entries are made on the matrix (Fig. 7.9), according to which criteria are better than, worse than, or the same as the datum. To increase an understanding of the specification, problems, and solutions, the matrix is rerun with the seemingly best concept or concepts assuming the role of datum. Once a clear indication is obtained of which idea should be chosen, the work is engineered to a higher level and in greater detail, and so leads to an expansion and revision of evaluation criteria. A new matrix is run and the process takes on a reiterative role that eventually converges to the best possible solution.

7.6 SUMMARY

Product development may be based upon either evolution, basic research, or innovation. In using one, or a combination of these three methods, a successful strategy is to treat the product, and the elements from which it is constructed, as two separate approaches each containing a number of tactical moves.

Scientific instruments, and process plant, are usually the result of innovation, but further development proceeds by evolution when 75 per cent of the ideas emanate from customers. A search for ideas should be based upon customers, competitors, and their suppliers.

Companies operating at levels of technology 3 and 4 differ with respect to the emphasis which they place on evolution and innovation. These companies form the majority in the United Kingdom and are characterized by large product portfolios. Two main sources of ideas are from the marketing and development staff. At levels of technology 5 and 6 ideas mainly come from research and development. Specially trained idea-generating groups are being increasingly used at those low levels of technology concerned with consumer products; creative techniques are also proving effective at levels of technology 3 and 4.

A description is given of two types of workshop which have been used in companies for the purpose of disclosing new business opportunities. The first is designed for six or seven of a company's executives or senior staff, and has two aims: the first to heighten a general awareness of the products and commercial situation of both a firm and its competitor; the second to take stock of all potential human and material resources. Eight procedures are used and range from technical audits to the construction of company scenarios.

The purpose of the second workshop is to enhance the existing entrepreneurial qualities of creative staff. The workshop is, again, designed for seven or eight people, two of whom should have attended the first. Three procedures are described whose purpose is to enable a three-dimensional business model to be constructed on which all present and possible future products are systematically located. At the intermediate and lower levels of technology a model should contain many hundreds of ideas, of which a small number of ideas should have sufficient merit to make the selection process straightforward.

The success of a new product is often critically dependent upon the excellence of its design, and a method is described by which a painstaking, systematic search for a new idea is combined with a technique for generating creative concepts. The choice of a solution, which is the most critical of all design problems, is aided by the construction of an evaluation matrix which allows cross comparison to be made of all proposals. The final stage is the revision of evaluation criteria and leads to a reiterative process aimed at giving an optimum design solution.

APPENDIX I
THE SYNECTICS PROCESS

Synectics is a generic label for a range of techniques, behaviours and meeting structures that have been identified as increasing the probability of success in invention and creative problem-solving. The methods have been established by taperecording and videotaping several thousand invention and problem-solving sessions over the last 25 years, and by experiment with alternative techniques in a wide variety of situations.

To the extent that the methods have been derived from observation of what people do when they are being successful, they can be considered as 'common sense', but they differ markedly from what is considered normal behaviour in a working environment (whether in business, industry, government or elsewhere).

One of the principal differences is that thought processes that normally go on privately, and sometimes subconsciously, are made explicit and public. By sharing the formative stages of their thinking, and consciously engaging in apparently irrelevant and speculative activities, the members of a synectics group greatly increase the range of source material that they access, and accelerate the discovery of novel connections and solutions.

To do so, synectics uses the brainstorming principle of suspending judgement, but extends it substantially so that it applies not only to ideas but problem statements, goals and wishes. Participants are encouraged to abandon the internal censoring of thoughts (which continues out of their conscious awareness even in a brainstorming session) by the use of a range of techniques involving metaphor, analogy, fantasy, visualization, association etc. The 'excursion' techniques provide distance from the problem and open up novel avenues of approach.

When it becomes necessary to reintroduce judgement, synectics does so in a way that maintains the constructive and emotionally safe climate created in the initial period of idea and concept generation. Developmental thinking is used to explore ideas which are emotionally attractive, though not yet feasible; all the potentially positive features of the ideas are identified, and the deficiencies are used to give the direction for improvement. In this way the elements of novelty are preserved while the idea is modified to make it feasible. (This process contrasts with conventional screening of ideas into 'good' and 'bad' after brainstorming, when novel ideas are likely to be screened out because they are not feasible).

To create and maintain the positive environment, the synectics session will normally be run by a skilled facilitator ('process controller') who takes no part in the content of the meeting. Responsibility for describing the problem, specifying the objectives of the session, selecting the ideas to be developed and evaluating ideas during the development phase, is in the hands of the problem owner ('client'), defined as the individual with decision-making and action responsibility. When, as in most organizations, more than one individual has some share in the decision-making, the model is extended to accommodate each of the problem owners, usually one at a time. The sequential handling of opinions and value judgements, combined with the positive climate and the generation of a wide range of options, makes synectics techniques particularly effective for the constructive resolution of conflict.

Many of the elements of the synectics processes have application outside the formal invention/problem solving meeting and can be used to develop a working style that is constructive, open-minded, tolerant of diversity and highly co-operative. The culture created in the synectics meeting can be extended to all the operations of the working group and it is in this kind of environment that innovation flourishes and change is managed productively.

The practice of videotaping and analysing the sessions continues for purposes of both research and teaching, and in this way the understanding of the invention process is deepened and the range of techniques is broadened.

A detailed description of the synectics process is given in Prince(23), and Nolan(24) explores ways in which it is possible to manage change to the advantage of everyone concerned.

APPENDIX II
A CLASSIFICATION OF IDEA-FORMATION METHODS

In 1971 the Battelle-Institut eV at Frankfurt initiated an extensive experimental research project on the field of creative idea-generation in industry. A group of 90 industrial sponsors was obtained; participating firms ranged widely in respect of size and business, and were located in six countries. A report of this work was publlished by Schlicksupp(25) in 1977 in which further research was suggested, particularly in the field of classification. In 1983 Geschka(14) from this same institution classified a large number of available techniques into the following four groups.

No.1: Intuitive association

This group includes methods which purport to assist the interchange and development of intuitive thinking among a group. Starting with classical brainstorming, nine other techniques are described, several of which were first devised by Battelle.

No.2: Intuitive confrontation

The purpose of this group is to trigger off new ideas in highly creative minds in the knowledge that inspiration often follows contact with events, objects, or thoughts correlated to the original problem. Five techniques are described, one of which is included in Appendix I.

No.3: Systematic variation methods

Here a systematic development of basic concepts is followed by individual solutions generated by a more creative process. Of these examples, one refers to a multidimensional morphological matrix, the use of which was described above (Chapter 6, p.162). Reference should be made to Geschka's paper for an account of progressive abstraction, which is a reiterative process used for companies seeking to diversify.

No.4: Systematic confrontation

This contrasts with no.2 by relying more upon a systematic approach than upon intuition. The first stage is to describe and formulate a problem or need, the second to identify the more important parameters, the third to construct a series of preliminary matrices, the fourth to seek alternative solutions, and the fifth to collate the ideas and information as they become available for the purpose of constructing a more informative matrix.

7.7 REFERENCES

1. Utterback, J. M. (1971). 'The process of innovation: a study of the origins and development of ideas for new scientific instruments', IEEE *Transactions on Engineering Management*, **November**.

2. Hippel, E. von. (1977). 'The Dominant role of the user in semi-conductors and electronic sub assembly process innovation', IEEE *Transactions on Engineering Management*, **EM-24**, No.2, 60–71.
3. Tushman, M. L., Moore, W. L. (1982). *Readings in the Management of Innovation*, Pitman Books Ltd., London, 131–150.
4. Parkinson, S. T. (1982). 'The role of the user in successful new product development', *R & D Management*, **12**, No.3, 123–31.
5. Wearden, T. (1981). 'Searching for new products - systematically', *Industrial Marketing Digest*, **6**, No.2, 153–8.
6. Lee, J. H. (1977). *Improved cost control by function cost analysis*, University of Southampton, pp.19–25.
7. Knowles, E. A. G. (1957/8). *Experimental design on the manufacture of certain plastic components*, Manchester Statistical Society, Group meeting, Industrial Group, pp.56–63.
8. Parker, R. C. (1982). *The Management of Innovation*, John Wiley & Sons, Chichester.
9. McCosh, A. M. and Kesztenbaum, M. (1978). 'A small-sample survey of the research financing process, with some implications for theory development', *R & D Management*, **8**, No.2, 65–74.
10. Dörfel, H. (1979). *Innovation in the Chemical Industry*, 12th International TNO Conference, Rotterdam, 22–23 February.
11. Gordon, W. J. J. (1961). *Synectics*, Harper & Row, New York.
12. Parker, R. C. (1975). 'Creativity: a case history', *Engineering*, **February, 215**. 126–30.
13. Holt, K. (1982). 'Idea generation—a key to successful management of change', *IHS-Journal*, **6**, 107–119.
14. Geschka, H. (1983). 'Creativity techniques in product planning and development. A view from West Germany', *R & D Management*, **13**, No.3, 169–183.
15. Robinson, S. J. Q., Hichens, R. E. and Wade, D. P. (1978). 'The directional policy matrix—tool for strategic planning', *Long Range Planning*, **11, June**. 8–15.
16. Fusfeld, A. R. (1978). 'How to put technology into corporate planning', *Innovation, Technology Review*, Massachusetts Institute of Technology, pp.51–5.
17. Zwicky, F. (1951). 'Tasks we face', *Journal of the American Rocket Society*, No.84, **March**.
18. Carson, J. W. (1974). 'Three-dimensional representation of company business and investigational activities', *R & D Management*, **5**, No.1, 35–40.
19. Carson, J. W. and Rickards, T. (1979). *Industrial New Product Development*, Gower Press.
20. Pugh, S. and Smith, D. G. (1976). *The Dangers of Design Methodology, European Design Research Conference—Changing Design*, Portsmouth Polytechnic, **April**, 1–19.
21. Pugh, S. (1982). *A new design - the ability to compete*, Design Policy Conference at Royal College of Art, London, 20–23 July, 1–25.

22. Pugh, S. (1981). *Concept selection - a method that works*, International Conference on Engineering Design, Rome, **March**, 1–10.
23. Prince, G. M. (1970). *The Practice of Creativity*, Macmillan Co. New York.
24. Nolan, V. (1981). *Open to Change*, MCB Publications, Bradford.
25. Schlicksupp, H. (1977). 'Idea-generation for industrial firms - report on an international investigation', *R & D Management*, 7, No.2, 61–9.

8

Research and Development at Both Medium and High Levels of Technology

8.1 PLANNING

The study referred to in the Introduction led to the publication of eight case histories(1) one of which (p.132–157) described a small company that was founded by four friends, and which had long enjoyed an open style of both management and good, informal staff relationships. The office of company chairman rotated between the four owners on a four-year cycle. However, uncertainties and anxieties born of the recession introduced a disruptive element which slowly eroded the team spirit. Not unexpectedly the main strains became centred on the marketing and technical directors. The former was aware that customers' loyalty was being impaired by broken promises, and the latter protested that the outcome of research and development was, and always would be, unpredictable.

By contrast, in a second company (reference 1, p. 100–130), morale was seen to be exceptionally high. Employees were familiar with company goals, were committed to their attainment, and there was general recognition of the high leadership qualities of the chief executive. Although the company had been forward-looking for over 30 years, as witnessed by their product range, a decision was made to meet the recession by innovating a new instrument for their main market. A specification was drawn up, and agreed, by the managers of both the marketing and technical departments, and detailed network plans were constructed for each stage of the development. Progress was monitored by liaison meetings in which membership and chairmanship were selected according to the state of development. The product was successful and was completed within the forecasted development time of three years.

These two case histories illustrate the importance of leadership and show that customer satisfaction is precariously balanced unless the product specification is detailed with meticulous care, planning is based on an appropriate methodology, and research and development is partitioned into a number of clearly defined stages. These factors will now be discussed in detail.

8.1.1 Product specifications

Once a proposal for a new product has been evaluated and selected, a great deal of information will have been collected about possible lines of development and the potential market. This will need to be incorporated into a detailed specification which should be stabilized as far as possible before significant expenditure is incurred. Creative discussions will usually disclose opportunities to lower cost targets, and to improve the original concept by the inclusion of new competitive attributes. It will be easier to surpass market expectations at higher levels of technology than at low, since research staff will possess unique knowledge on the potential fruits of a new technology. At intermediate levels of technology, too, technological information will, normally, reside with manufacturers rather than with users. It was the former, for example, who recognized the possibility of designing watches which were antimagnetic, shockproof, and waterproof.

The compilation of a specification will often alert a design team to the need to know more about the usage to which a product is likely to be subjected. This is particularly true for components supplied for incorporation in large industrial devices, when it is more usual for a specification to be supplied by the customer. Specifications issued by industrial users are often misleading because they pay insufficient attention to usage patterns: overload tests may be based merely upon a higher performance than was obtainable with previous products and may be irrelevant to needs. In those industries whose products contain a high proportion of bought-in components the specifications for which would entail field research, suppliers can secure an advantage by carrying out usage research for their customer. In illustration of the way a specification can be unrealistically upgraded, an engineering journal, in the 1960s, published a specification for a coach brake which was equivalent to the coach being driven at 13.50 Km/hr down a 1 in 10 gradient for five hours.

Responsibility for drawing up a product specification should be vested in the chief executive of either marketing or sales, and contributions should be sought from staff likely to be directly or indirectly involved. For industrial products a proposed specification should always be discussed with the customer before it is finalized.

The higher the quality of a specification the greater the likelihood of achieving and maintaining customer satisfaction, and lower the risk of wasting development expenditure.

8.1.2 Network

A critical factor in the DCF method of project evaluation (page 222) is the launching date of the product. It is recommended that the work be subdivided into a series of parallel events or activities in order to facilitate completion of the development on time and at the agreed cost. Figure 8.1 shows a block diagram that was constructed to decide between two proposed production methods, based on three different concepts, for developing a friction material.

Although a convenient step, when planning a development programme, is to construct a block diagram, it should be converted into one of the more formal networks as soon as sufficient data has been collected. These networks have a number of advantages. They show not only when every activity should be started, but when they should be completed. Although, in most cases, estimated durations may be exceeded, this will not be so for one route through the network, termed the critical path.

The PERT type of planning network, though widely used for large mechanical and civil engineering projects, is not the most suitable for laboratory development work. Instead, the precedent diagram, published by Roy(2) in 1959, is preferred on the grounds that its logic is closely akin to the mental processes involved in development work. It has three advantages over PERT: it does not require 'ladders', the only 'dummy' is the final activity known as 'start-up', and activities are shown at the nodes and not on the arrows. It is simple to construct, easy to read, and can be quickly updated. Figure 8.2 shows the nomenclature and a simple example of a precedent type of network, and Fig. 8.3 shows the translation of the block diagram of Fig. 8.1 to this form.

The construction of a precedent type network combines the functions of planning and monitoring, and those activities which dominate important outcomes should be carefully watched in order that corrective action can be taken should progress falter. A carefully constructed network not only helps to assign responsibility but also aids delegation. Network activities can be listed and used as the agenda for the progress committees discussed on page 195.

Because every activity in a network has its associated material and human resources, it can be used to plan the best way in which activities may be sequenced in order to avoid delays caused by lack of manpower, late deliveries of supplies, and service queues. This use of precedent networks can be effective for work which, at first inspection, appears to be mundane and incapable of improvement. Parker and Sabberwal(3) describe how a network was used to plan the preparation of frictional samples, their curing in ovens and the final measurements of their physical properties. Even in this repetitive work a critical path was indicated and so provided an opportunity for levelling out the demand for resources. The final stages of this work are shown in Fig. 8.4, from which it is seen that, whereas in the first plan no less than nine assistants would have been allocated in weeks 27 and 28 to measure the physical properties and analyse the results, in an amended arrangement only two were required at any one time.

Research, at the highest technological levels, operates at the frontiers of science, and it is rarely possible to forecast when, and if, a successful outcome will be obtained. The above planning methods are only relevant, therefore, at the later stages of development. This type of work will normally have expectations of both high benefits and risks, and may involve unknown factors in the production processes. For these reasons it is not always appropriate to construct planning diagrams.

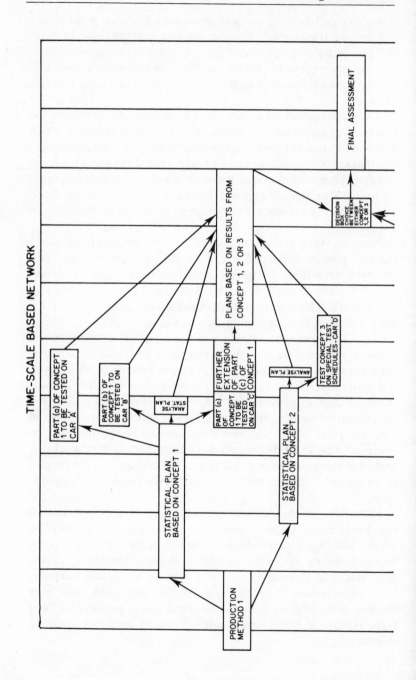

TIME-SCALE BASED NETWORK

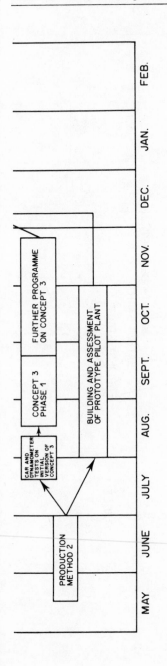

Figure 8.1 Simple block diagram

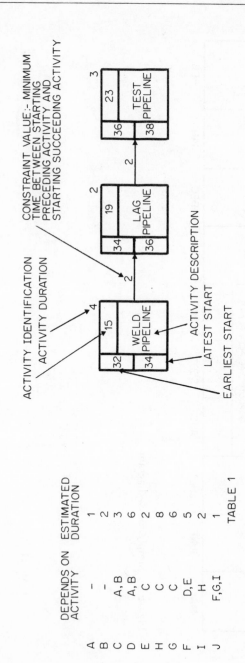

TABLE 1

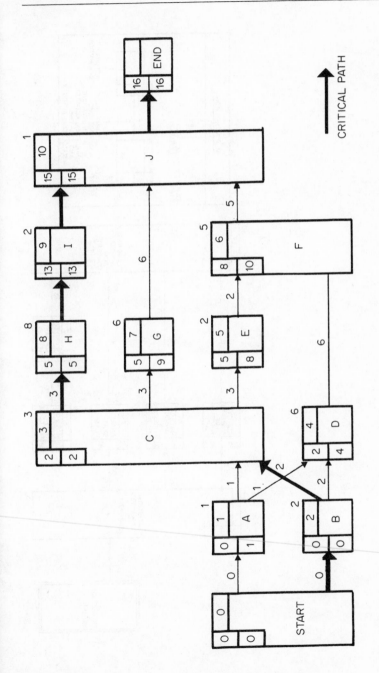

Figure 8.2 A precedent network

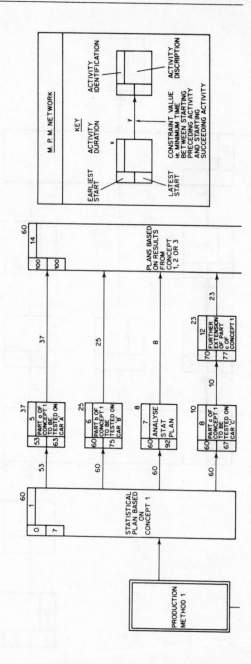

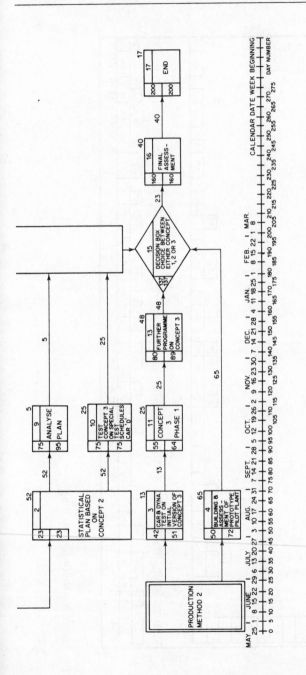

Figure 8.3 Conversion of block diagram (Figure 8.1) to precedent network

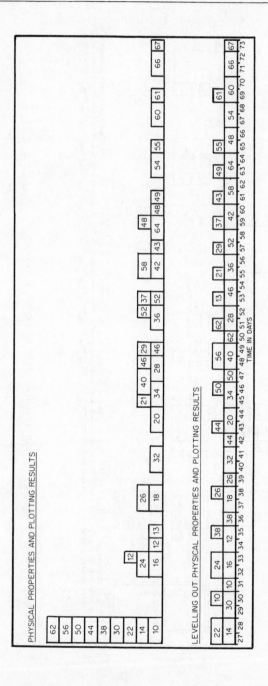

Figure 8.4 Resource levelling

8.1.3 Five stages of development

Projects, apart from the small number associated with high technological levels, should be so managed that prototype production samples, which are normally submitted for customer approval in advance of sanctioning full production, must be matched by the subsequent production version. The loss of confidence by the customer if this is not so is extremely difficult to regain. The recommended procedure to be used is to divide the sales development and production responsibilities into five separate stages as illustrated in Table 8.1.

Table 8.1   Assigning responsibilities when developing a product

Stage	Description	Responsibility
I	Acceptance that a product submitted by R & D to sales is suitable for provisional submission to a customer	Sales
II	Initial prototype production controlled by R & D for submission to customer in limited quantities	R & D
III	Production quantities manufactured on factory plan under supervision of R & D staff	R & D manufacturing
IV	Production quantities manufactured by factory employees and checked for quality by R & D staff for an agreed initial period	Manufacturing
IIx	Provision for submission to customer of substantial quantities prior to completion of prototype development	Sales and R & D

Stage I involves development work carried to the point at which a sales department accepts that it conforms with the product specification. It is regarded as complete when the R & D department supplies sales with data that characterizes the product, and additionally, gives an early indication of product costs and suitability of the process for efficient manufacture.

In stage II prototype samples of the product are manufactured by research and development staff, either on factory plant, or on identical plant in a research experimental production department, for submission to a number of representative customers. The quantity produced must be sufficient to highlight possible manufacturing problems, and should permit assessments to be made of quality control techniques, reject levels, and costs. All relevant data should be formally handed to sales and works divisions. The stage is not complete until a sufficient number of customers has given provisional approval to the submitted samples.

The purpose of stages III and IV is to transfer responsibility for the production process from research and development to the manufacturing division. In stage III customers' orders are accepted and progressed in the factory, but under the supervision of technical staff responsible for design and development whose task is to ensure that the product can be efficiently manufactured to a defined standard. In stage IV production is carried out entirely by the factory, but full responsibility is not accepted until the end of an agreed period of some months. This ensures that the R & D staff are satisfied with the maintenance of quality, and the works staff are confident that the product and process are compatible. At the end of the agreed period the production director accepts total responsibility for the product and may not involve sales or research and development divisions unless evidence is produced of unavoidable, unexpected, and possibly unexplained causes, responsible for unacceptable changes to the product.

Stage IIx is introduced to cover the dilemma that arises when a customer demands production quantities of a product during stage III. It may lead to serious difficulties when production is transferred to research supervision, because it will inevitably place a heavy burden on their resources. It should only be agreed to in exceptional situations.

8.2 ORGANIZATION

When the first of the two companies, referred to in paragraph 8.1, realized that the business situation was rapidly worsening, and their open style of management was under strain, the four owners embarked upon a frank discussion of the company problems. They recognized a need for an improved management structure which would strengthen controls and prevent duplication of work by distinguishing between staff and line function. They further decided to appoint a managing director to set targets, measure progress, and accept responsibility for corporal decision. The office of chairman continued to be shared among the four founders. The problem of introducing a degree of formality often occurs during early stages of company growth (see Chapter 2) and the design of a new organization must aim to retain those good relationships which are characteristic of small companies. No one pattern can be recommended because directors face contrasting situations and there will be wide variations in the size of proposed departments, in the age and experience of staff, and in the dynamism of the companies.

Based on experience obtained by changing organizations to adjust to a 30-fold increase in staff over two decades, a few basic principles will be described that are believed to be applicable to the difficult situations that occur at intermediate levels of technology. Figure 8.5 illustrates the main activities of a research

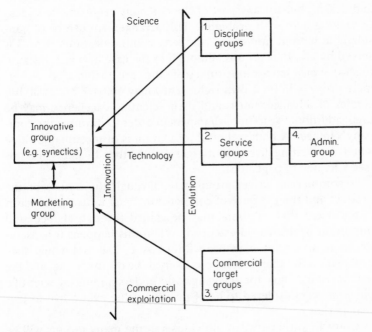

Figure 8.5 An organization scheme

and development division operating at levels of technology 3 and 4. Box 3 represents the important activities. Designated 'commercial target groups' it is concerned with research, design and development skills in producing improved or new products, and is therefore one upon which the company's future ultimately depends. It also signals, unambiguously, that the purpose of the division is to serve sales. There may have to be many parallel groups in order to cover several classes of product and, possibly, a number of manufacturing processes.

If the R D & D scope is wide, box 1 may represent many functions: for example, engineering, mathematics, physics, metrology, stress analysis, and polymer chemistry. Box 4 will be responsible to the research and development director for all administration, including information services, and box 2 will comprise testing, analyses, and similar services. At the left-hand side of the figure is shown an innovative group, closely linked with marketing. It can act as a resource group not only for research development and design, but for any part of the company requiring new ideas or seeking help with problems. This arrangement can be of considerable help in improving company communications. Box 3 is served by all other groups, although in the case of box 1, because this work may have a long time scale they are encouraged to originate projects in their own right. They may embark on feasibility studies of advanced ideas and their scientific excellence may be used, additionally, to publish papers in order to improve the company's image. A high technical reputation is of inestimable value when negotiating licences, mergers, and other business linkups (see Chapter 5, page 106).

A commercial target group may, through the nature of the product and the pressure of competition, need to use empiricism to an extent that solutions may be gained through chance, and important problems side-stepped. When this happens it is useful to design a two-tier structure within box 1. The first would seek explanation for chance solutions likely to be of future use, and the second would look for relevant basic understanding of scientific phenomena considered to be of possible value over a five to ten-year period.

Co-ordination between individuals in the many groups will be effected largely by the system adopted for reporting and planning, but if activities are carefully specified each group will retain the degree of autonomy which is necessary for maintaining morale. Service groups, in particular, must be made aware that their contributions are as vital to success as are the functions that have greater visibility to other parts of the organization. Cheng(4) in a study of coordination, uses the term 'reciprocal groups' when the output of each task becomes the input for others. He notes that an arrangement of this kind involves a social process of mutual adjustment if morale and cooperation are to be sustained at a high level, and individuals should negotiate expectations of each other's contribution, and agree on adjustments which have to be made.

When designing an organization it is inadvisable to draw a too sharp distinction between project and disciplinary orientated developments. If a group needs to attain mastery of complex theoretical concepts, or requires seasoned professionals in highly specialized fields, then members should generally remain together and so inspire each other and keep abreast of progress. Again, if breadth is required in a technological subject, for example, a complex engineering or chemical process, then staff are likely to benefit from working in a team over a long period. Commercial target groups, on the other hand, will generally consist of a few prominent members, expert in sales needs and their particular technology, but may from time to time be supplemented by other staff with appropriate skills. The many personal factors which need to be considered have been fully discussed by Pearson, Payne and Gunz(5). Figure 8.6 shows a flexible arrangement whereby commercial target teams may be loaned members from disciplinary and service units. Transfers should rarely occur and the periods should not exceed a few months at a time. Should it be found that a particular discipline has been denuded of the majority of its members, then those remaining should be encouraged to take up long-term advanced projects.

Where multidiscipline tasks are involved the aim should be to avoid interdisciplinary arrangements wherein input from individuals has to be integrated by others, and should adopt cross-disciplinary teams which integrate their own inputs.

In contrast to the above, companies operating at levels of technology 6, and possibly 5, will largely comprise groups represented by box 1. They will carry out their own commercial target functions. A few service groups may be required, but there will rarely be a need to set up administrative or separate innovative groups.

Should a research director sense that his division is failing to achieve its potential, he may decide that a new organizational structure is indicated. This is likely to be a major source of uncertainty, since changes in responsibilities will please some and dismay others. While an ideal target must be sought, it is wise to

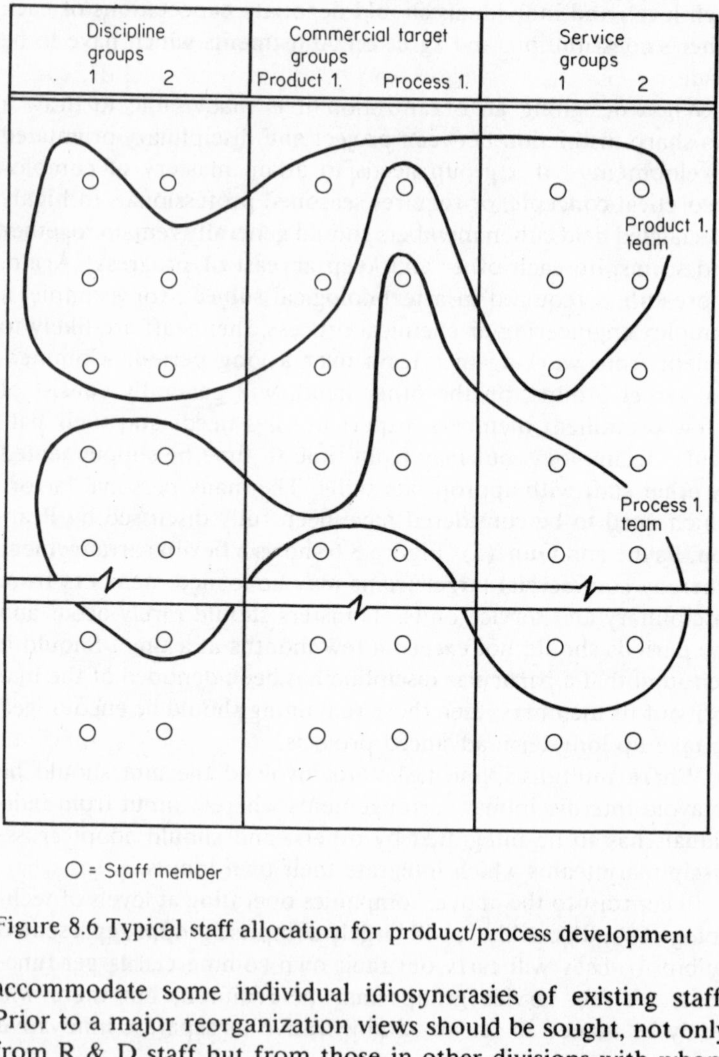

O = Staff member

Figure 8.6 Typical staff allocation for product/process development

accommodate some individual idiosyncrasies of existing staff. Prior to a major reorganization views should be sought, not only from R & D staff but from those in other divisions with whom close cooperation is maintained. The final proposal should be circulated before implementation, and the director should be available for discussion with any individual who is experiencing concern.

Company environments vary so widely that Fig. 8.5 cannot apply to all situations. For example, in many industries there is a strong argument for merging research development and design, for both the product and manufacturing processes. This may be vital in those cases where the properties of a product depend, in part, upon the production process. It further confers the advantage that staff can be more easily assigned to different tasks and can, in particular, accept responsibility for the initiation as well as the later stages of product development. Johne(6) in investigating a number of organizations in manufacturing firms involved in technology, observed that organization structures tend to be loose in the initiation phase of a project, and become tighter in the implementation stage. This again indicates that a good R & D organization should have an inbuilt capacity for flexibility.

8.2.1 Committee structure

One consequence of the diversity associated with medium levels of technology is the difficulty of delegating responsibility for a project to an individual who can combine both the functions of power and know-how (see Chapter 2, p.35). A practical solution is to use a committee chaired, in turn, by 'manager-leaders' who have had close association with pre-project work. The first chairman would probably be the marketing manager who drafted the original request for a project, whose initial task should, as we have seen, be to seek information to enable the product specification to be defined. Committee members would be restricted to staff whose contributions were deemed essential. For the next phase the marketing manager may leave the committee temporarily, when his place might be taken by the technical manager whose responsibility would have been to develop the product concept. On completion of his responsibility, work could proceed to developing a prototype design, when the next most suitable candidate for leadership would be a member of the manufacturing staff. In a final stage, the marketing manager may again take the chair in order to plan the product launch.

In a company in which cooperation is good, changes in manager-leadership need not be formal but can be carried out according to group consensus. The author has frequently noticed that it is the manager-leader who suggests a change in chairmanship. As manager-leaders rotate, so will the nature of the tasks and, hence, the composition of the committee. The committee should report to an executive director who, in the capacity of sponsor to the project, should attend all meetings, so that the board can be acquainted with progress and give the work the necessary support.

The committee's function will be to monitor progress, amend networks as required, air difficulties, indicate possible solutions, and suggest sources of external help. The precise objectives of the committee should be explicitly stated beforehand. If the purpose of the meeting is to monitor the project, members will need to be armed with up-to-date information, and be prepared to arrange additional coordination and to take decisions. If, on the other hand, the purpose is to modify plans, either because the specification has been altered or technical difficulties have arisen, the emphasis will be on know-how and policy. Since the committee's work will, eventually, directly affect company profits, it will be important to review, continually, every aspect of the project. It could be that amendments to the specification are necessary to cover new eventualities caused by environmental changes, unexpected competitor activities, or difficulties with supplies of raw materials or components.

8.2.2 Venture management

A venture group is an entrepreneurial team formed within an existing business for the purpose of establishing a new growth enterprise. Management of a venture is vested in one individual. It may begin as an in house activity but is not part of a research and development function.

During the late 1960s and early 1970s Dunn(7) reported that while almost half the top 100 US companies had venture groups, their casualties became high, and popularity has waned.

Among an exceptionally large number of publications on venture management a paper by Tharby(8) of INCO is one of the few comprehensive reports by a practitioner. The company's history is one of dominating the nickel market and was sustained by developing new uses for the metal. A major contribution to its success was two large research laboratories, situated in the United States and United Kingdom respectively.

Competition evolved as a serious threat in the mid-1960s, and INCO's response was to adopt a policy of diversification of which internal venture development was a major contributor. It was initiated within laboratories that were given a remit to develop ideas dissociated from the mainstream of the business. The aim was to use areas of technology of which the laboratory had some knowledge, supplement it where appropriate with outside help, and to look for products with acceptable entry costs even though risks were high and time scales long. The market objective was to develop unique, outstanding products, secure a high market share, and achieve product earnings which were significant to the company.

Two small autonomous subsidiaries were formed to act as a nursery to new ventures until they evolved as self-supporting enterprises. The first, MPD Technology Corporation, was launched in the United States in 1976 and was followed by MPD Technology Limited in the United Kingdom in 1978. Emphasis was placed upon the importance of keeping communication lines short, and retaining a capacity to respond rapidly to changing situations. Individual ventures had the advantage of sharing the cost of administration and other support functions. Five specific areas of venture technology emerged, namely: electrochemistry directed at production of nickel foil and surface treatments; special plastics designed to be directly electroplateable; hydrogen-related energy; special metal powders and related products; and metal forming.

All venture managers were appointed from research and development staff, the majority of whom were closely associated with the original concept. Other team members were often external appointments. On joining the venture team, staff were warned that should the venture fail no guarantee of re-employment could be given, and a project was required to be demonstrably viable within a stipulated period. The motivation was regarded as a long-term opportunity for personal growth and qualification. It was recognized by the companies that, on average, a new venture needed eight years to become profitable, and 12 to perform as a mature business.

The INCO strategy of forming a separate company to house venture groups is unusual, but in other respects it fulfils the first aim of venture management which is to operate outside traditional, authoritarian structures in which bureaucracy may hamper development, and existing staff often regard new products as a threat to existing business.

The first criterion for a successful venture concept is that it must be capable of being developed with no other resources than those normally available to an entrepreneur setting up a new business. During a long period of directing research into friction material products, the author recalls only one idea which did not require a wide range of costly resources. Though then unaware of new venture concepts, this lone idea was developed in an empty mill, some 16 Km distant, to allow freedom from overriding priorities for ongoing, more conventional developments. The manager eventually became the general manager of a new factory, and was able to combine the motivation, of the kind experienced by owners of new businesses, with an opportunity to gain advice and limited resources from a large organization.

Based on a consensus of published material a venture team should have above-average ability, be young, and comprise research scientists, technologists, a market-orientated researcher, a design engineer, and a production engineer. Its manager should be aggressive and ambitious, be keen on empire-building, have realistic expectations, and a good track record of past success. The team should be sponsored by an influential member of the board who has earned the respect of his colleagues.

Venture groups have a role in growth, but are not a cure for an ailing research and development department. When forming a venture group it is essential to record its functions and likely time-scale with great care, and in meticulous detail, and to make certain that it is agreed by all executives.

One crucial task of a venture manager is the initiation and development of a professional business plan, which must indicate substantial benefits if funding by the company is made conditional on the manager obtaining equity funding from external sources. Jones and Oakly(9) noted that in the United States it is common for all members of a venture team to invest substantial amounts of their own funds, and consequently their endeavours for success and reward are of a high order.

With the exception of companies such as General Electric and Exxon, venture management has not sustained its early reputation for success. Crises have changed company policies within the time-scale of a typical venture group, and world recession has destroyed promising markets. In a comprehensive literature survey Burrows(10) cites many causes of failure, most of which can be attributed to Schon's dynamic conservatism(11). Opposition from within a company is not unusual and is attributed to resentment of what is regarded as an elite group, and to the fear of impingement of territories. Members of a venture group see things differently, and query whether they have a passport for success or an exit pass.

A venture group cannot succeed without enduring corporate support. Its formation should reflect a firm's commitment to growth, and never be merely a short-term reaction to a downturn in business. It is significant that whereas INCO believes that a new venture requires 12 years to mature, Dunn (p.196) reported in 1977 a high failure rate for ventures, none of which had run for seven years. The widely reported failure of venture teams reflects the reluctance of industry to plan well ahead.

8.3 LEADERSHIP

Good ideas are translated into successful products through the exercise of leadership. The basic qualities common to good leadership are discussed in Chapter 2, p.38, but there are additional attributes relevant to research and development activities, and of these some will apply to laboratories concerned with high technology and others to low technology products.

The high technology company, whose strategy is to maintain a flow of unique technical innovations, will have an imaginative management of the highest calibre, and its laboratory will be staffed mainly by scientists and technologists predominant in their field. To keep ahead of competition a laboratory atmosphere must be created which can not only attract the best brains but retain them. To discharge this, and other functions, the head of research should have director status and be an eminent scientist or technologist who has a natural empathy with conditions most likely to motivate his staff. For example, rather than accept the lines of authority normally found in industry, he may prefer to adopt the 'acceptance theory' which, of the three theories of authority reviewed by Van Fleet(12) is probably best suited to scientific staff. In this an individual behaves as if his assent to a request is conditional on his full and willing acceptance of authority. Instructions must be understood, must be seen to meet the purpose of the organization and the individual's personal interests, and be thought capable of achievement. However, in practice, it is not essential for all requests to be accepted willingly, since there are always areas of indifference in which orders are acted on without questioning. This atmosphere encourages adaptation, and the setting up of new states of equilibrium, and it is, therefore, particularly appropriate to the unstable environment which must be characteristic of vigorous research and development.

To establish the concept of acceptance authority, participating decisions must always be encouraged, and especially where ego and self-actualization needs are involved. The individuals in an organization of this kind will need to be capable of sustaining learning, adapting, and evolving. It has been observed that although scientists often state a wish for autonomy in choice of research topics, given a choice, they opt for group discussion in which they have equal representation.

The research director must establish contact with scientific centres of excellence, and familiarize himself with sources of know-how which may be needed later by his staff. He should encourage scientific publications in order to widen contact with others working in the same or associated scientific discipline. A major responsibility, of his status as a director, will be to ensure that progress is not hindered by reason of inadequate resources.

At medium levels of technology an R & D director has to deal with a battery of complex situations because the product portfolio is likely to comprise dissimilar products manufactured by many different processes. Unlike his counterpart in a high technology environment, he will need considerable skill in coordinating widely diverse projects, and will have to exercise considerable administrative ability to maintain momentum of competing demands. Because innovation coexists with evolution direction will necessarily have to be flexible. Though his scientific and technical ability must be sufficient for him to recognize which of the many demands from sales and works merit priority, he will rarely have either the time or opportunity to exercise leadership by virtue of his professional attainments.

The better an R & D director understands the motivation of those with whom he has to deal, outside his division, the better will he direct his staff. For example, Pearson (personal communication) has remarked upon four ways in which marketing and R & D staff differ. The former forget failures and quickly embrace new enthusiasms, and are committed to company profitability. The latter are prone to remember failures because they live with and accept uncertainty, and their commitment is to projects. The former also tend to spend the larger sum when innovating new products.

Good communications will clearly be essential in such varied and changing elements, both within and outside the company, and should the director be so fortunate as to recognize staff acting in the role of gatekeepers (see Chapter 4, p.88) he should arrange for them to work in a readily accessible part of the laboratory.

Though the tactics of research directors may vary within different companies, a common aim must be to encourage originality, receptivity to change, and a willingness to communicate. The ultimate aim will be to encourage a spirit in which the loyalty of individual groups merges to give a divisional *esprit de corps*, which, in turn, is sublimated to a consciousness of company purpose.

8.4 SUMMARY

When drawing up product specifications it is important to know the stresses to which products are likely to be subjected, and it is often advisable for a supplier to carry out user research for a customer. The responsibility for drawing up these should be vested in both the chief marketing and sales executives and, ideally, the goal should be a product which surpasses the expectations of the user.

To ensure that development is completed on time, planning networks should be constructed. A precedent type is recommended, for this serves both the planning and monitoring activities. It can be easily updated, and can be used to optimize the allocation of resources. To lessen the risk of losing customer goodwill through failure to reproduce prototype characteristics in full-scale manufacture, development should be subdivided into five activities when responsibility for quality is formally transferred between the manager at each stage.

The organization of a research and development division will vary from company to company, but there are a few basic principles which should be adopted. The function of the main department is to serve sales and thereby the customer. Other departments are regarded as fulfilling a service role and are, therefore, similarly directed towards achieving customer satisfaction. An organization should be flexible and the control of a project will become increasingly more formal as the work passes through its five development stages.

Committees are required to monitor progress, amend networks as required, and ensure that the project completion date is met. The committee should be chaired by the manager of the appropriate development stage, and the point of change should be decided by group consensus.

Venture groups are examined as a possible way of using entrepreneurial teams, and reference is made to possible causes of their success and failure.

At high levels of technology R & D may employ a greater number of eminent specialists than in any other company function, and this requires a high quality of leadership if the most able are to be recruited and retained.

8.5 REFERENCES

1. Parker, R. C. (1982). *The Management of Innovation*, John Wiley & Sons, Chichester.
2. Roy, B. (1959). 'Contribution de la théorie des graphes à l'ètude de certains problèmes linéaires', *C. R. Académie Sciences*, **T.248**, 24–37.
3. Parker, R. C. and Sabberwal, A. J. P. (1971). 'Controlling R & D projects by networks', *R & D Management*, **1**, 3, 147–53.
4. Cheng, J. L. C. (1979). 'A study of co-ordination in three research settings.' The management of research groups, *R & D Management*, **9**, Special Issue, 21–20.
5. Pearson, A. W., Payne, R. L. and Gunz, H. P. (1981). Communication, co-ordination and leadership in Inter-disciplinary research', Associate Membership Scheme paper, Manchester Business School, pp.1–21.

6. Johne, F. A. (1983). *The Organisation of Product Innovation in High Technology Manufacturing Firms*, City University Business School, Working Paper No.50.
7. Dunn, Dan T., Jr. (1977). 'The rise and fall of ten venture groups', *Business Horizons*, **October**, 32–41.
8. Tharby, R. H. (1980). *Venture Group Management—Problems and Opportunities*, Society of Chemical Industry Joint Management Group, Symposium 'Seeding Innovation', **20 March**, 1–18.
9. Jones, W. H. and Oakly, M. H. (1979). 'A case study in venture management', *OMEGA*, **7**, No.1, 9–13.
10. Burrows, B. C. (1982). 'Venture management—success or failure?', *Long Range Planning*, **15**, 6, 84–99.
11. Schon, D. A. (1973). *Beyond the Stable State*. Pelican Books, Harmondsworth.
12. Van Fleet, D. (1973). 'The need—hierarchy and theories of authority', *Human Relations*, **26**, No. 5, 567–580.

9

The Funding of Growth and the Selection of Ideas

9.1 CONSTRUCTING A SOUND BASE FOR GROWTH

The primary aim of a company should be to generate sufficient funds to ensure continuity. A successful interaction between current and future activities is a crucial factor in both survival and growth, and brief reference will be made to a few basic problems which regularly resurface when seeking to run a profitable operation.

A feature which characterizes inefficient businesses is the frequency with which unplanned activities occur. Because panics seem to arise without warning, and vary in nature, they are too often accepted as inevitable. Unfortunately they can cause insidious, long-term harm. If, for example, the promised launch date of a new product is endangered, both marketing and sales may pressurize development and works teams, who will, in turn, be tempted to adopt short cuts, with deleterious consequences both to design and production. Inviolate priorities cannot be sustained in the face of recurrent panics.

Once the pressures of immediate problems are lessened, executive staff can pause every few years and make a thorough audit of their business operation. A routine is described on page 155 in which the basic elements of a business are examined to ensure that they contribute in a coherent manner to the total company performance. The audit should also include a comparison between the company's and the competitors' activities.

A comprehensive review should show where effort, aimed at improved efficiency, may be most effectively directed. It could be towards lowering production costs, reducing the manufacturing cycle, shortening delivery times, and improving quality control, reducing work in progress or reducing the stock of raw materials.

Care will always be needed when introducing economies. It may often be advantageous to retain dual sourcing of critical raw materials, even at higher cost, and to carry large stocks of supplies from countries threatened by political instability.

Growth based on new products is particularly difficult when the existing range is large. Effort tends to be too thinly spread, and panics multiply. This characteristic is commonly experienced by companies operating at levels of technology 3 and 4. Some companies try to eliminate products but make little progress due to customer resistance, and others adopt group technology, or a flexible manufacturing system based on computer control, to minimize production difficulties caused by a proliferation of part numbers. A partial solution lies in Sumner's suggestion of always designing for production (personal communication). He refers to an article(1) which compares the experience of a United Kingdom and a US company who were in the same line of business, and comparable in all respects other than in profitability. The manhour element in the US company ranged between 70 to 48 per cent of that required in the United Kingdom company. Sumner makes two recommendations:

(1) Design should aim at a minimum number of
    (a) products for the market;
    (b) parts for the range of products;
    (c) processes and materials to suit the product range.
(2) A board should lay down a policy whereby
    (a) no new parts or subassemblies should replace satisfactory current ones;
    (b) designs should be for future common use;
    (c) the number of parts should be controlled by creative techniques.

Sumner underlines these approaches by citing two car manufacturers, one of whom uses two door handles, and the other 20 for fewer models. Small(2) has found that over 50 per cent of unit costs are attributable to materials, bought-out components, and inventories, of which a half may be caused by obsolescent methods and practices. Too much attention tends to be centred on direct labour costs which rarely account for more than 10 per cent of unit costs.

A programme designed to establish an efficient business will heighten managers' awareness of available new technologies such as computer-aided design and manufacturing systems, microelectronics, robotics, and automatic warehouse systems; and there will be a temptation to introduce change in a piecemeal manner. This will only be worthwhile if there is both spare technical and financial support.

It cannot be too strongly stressed that unless action is taken to avoid recurring production problems, product development cannot proceed smoothly. It is recommended that companies plan, on a continuous basis, a major redesign of their plant every few years. For manufacturing industries, at medium levels of technology, this could be five years, and a longer period for higher levels. This strategy is necessary because work on existing processes and products inevitably reaches the point of diminishing returns, when development teams introduce more frequent modifications of ever-decreasing significance. The effect on production and, hence, on development departments, can be catastrophic: work study and time study staff become overloaded, union–management relationships deteriorate, and the works director may become antagonistic to change. A similarly adverse impression is created on customers who, wishing to retain good relationships with suppliers, accept modifications which they believe are not worthwhile.

The introduction of planned updating of plant gives the design and development teams greater opportunities and challenge, ensures that productivity and production costs will be kept under continual review, and maintains a lead over competitors. When a manufacturer has several major processes the designated completion dates must be evenly spaced out, so that one or more years elapse between the introduction of change. Without an orderly approach of this kind, staff responsible for innovation lack a secure process or manufacturing base on which to design new products.

9.2 FINANCING GROWTH—A CORPORATE VIEW

Unless companies have prepared for growth over a number of years they will not find it easy to raise sufficient funds to implement their selected ideas for improving the product range. Many companies do little more than make an annual allocation from their R & D budget to cover new product development. This, however, is of little practical value unless all divisional budgets are increased in sympathy. The cost of a new product is likely to involve patents, marketing, startups, tooling, designing, and industrial engineering, which, together, may exceed R & D expenditure by a factor of 3.

New high technology companies will include growth as a main target, when future funding will be a major consideration. By contrast a large number of subsidiary companies responded to the recession of the late 1970s by retrenchment, and many remain unaware that a drive towards efficient company operation can coexist with new product development.

There are many paths to growth but, irrespective of which are chosen, work should not begin until full consideration has been given to possible sources of finance.

There are four basic options: to generate funds internally, obtain equity finance from merchant banks or other financial institutions; raise funds from existing stockholders through rights issues, and to raise loans or make credit arrangements with clearing banks, venture capital groups, and similar bodies. Equity finance is attractive to the owners of businesses on two counts. Payments are not called for until a degree of success is attained, and, in the absence of fixed interest and security, a business is unlikely to be lost altogether. Equities do, however, carry voting rights, and control of a company may be lost more easily than with a loan.

As we have already seen, advanced technology means increasing both development costs and lead times and this accentuates the problem of finance. A particular need is for easier access to capital markets which would have a longer-term outlook. It has been argued that financial institutions should render the intervention of government institutions unnecessary in this field, and banks should provide equity finance, loan finance, and credit facilities.

The provision of finance for growth is a major responsibility of management, but is among the most difficult because uncertainties associated with new products are compounded by ignorance of competitors' actions. Whatever the problems of a particular company, ideas may be gained by looking at studies which have been carried out on macroeconomic aspects of growth.

Jones(3–5) Cox(6,7) and Cox and Kriegbaum(8) have used the concept of value added as an analytical tool for examining a company's potential growth where value added is the difference between a company's output, expressed as the value of goods and services produced, and the input or cost of raw materials, energy used in production, and services bought in. The value added must, in the long term, be sufficient to cover the cost of both current and future activies.

Jones, using data supplied by the Central Statistical Office, examined the manner in which UK and Japanese companies distributed their value added, and has compared their industrial performance. Figure 9.1 shows (a) the distribution of net added value per employee in the UK industry for 1976; (b) the comparable figures for Japanese manufacturing industry for the same year; and (c) the UK figures for 1973. Industrial experience has shown that the 10.5 per cent of added value retained in a company (Figure 9.1a) is inadequate to maintain solvency, whilst financiers will not be encouraged to underwrite expansion for a return as low as 1.9 per cent of added value. Comparison of Figure 9.1b with 9.1c shows a disquieting and adverse trend, namely, an increase in the government's share from 24.7 per cent to 31.4 per cent, which, if it continues, must exacerbate the problems of financing a company's future. Based on a study of United Kingdom data for 1977 Jones argued that United Kingdom industry should approach more closely the Japanese performance, and aim to retain 20 per cent in the business, while target figures for employees' take-home pay, government taxes and charges, and financiers should be 40, 20, and 20 per cent respectively. He concluded that, in view of our long period of zero or negative growth, and the political unacceptability of reducing employees' pay, only one solution remained—a substantial blood transfusion in the way of cheap capital. Although future changes in personal taxation, corporation tax, and interest rates will modify Jones' target figures, his method of analysis is sound, practical, and should be more widely used. It should be mentioned that many believe Japan's success is partly attributable to their borrowing at a high rate of gearing.

Distribution of net added value in UK industry in 1976 (per employee)

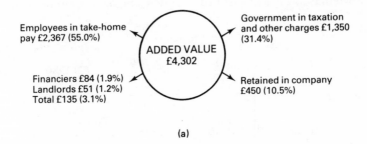

(a)

Distribution of net added value in Japanese manufacturing industry (per employee)

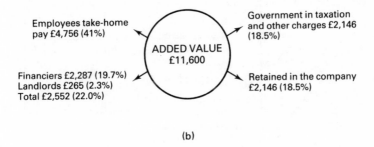

(b)

Distribution of net added value in UK industry in 1973 (per employee)

(c)

Figure 9.1 Distribution of net added value in UK and
Japanese industries

A significant proportion of the 20 per cent retained in a business should be invested in production plant, but Sumner has shown that this is only justified if it gives a higher output per man.

Capital Productivity

	RISING	STATIC	FALLING
Rising	G Excellent	G Good	Y Poor
Static	G Good	Y Stagnant	R Bad
Falling	Y Poor	R Bad	R Disastrous

Labour Productivity

Figure 9.2 Capital and labour productivity

Figure 9.2 shows a matrix which he used to illustrate the need to link capital productivity (high value added per unit of capital employed) with labour productivity (high value added per pound of employees' take-home pay). Jones' work indicates that United Kingdom industries use capital and assets as efficiently as any country in the world, but the level of investment in assets is too low to allow employees to achieve satisfactory outputs.

Cox's approach has been concerned more to show how a business should be run to give the highest possible value added, with particular reference to both materials and employee costs. By collecting and interpreting statistical data over a period of 30 years, she has compared current costs and the funding of investments within four United Kingdom and German industries: mechanical engineering, motor vehicles, chemicals, and textiles.

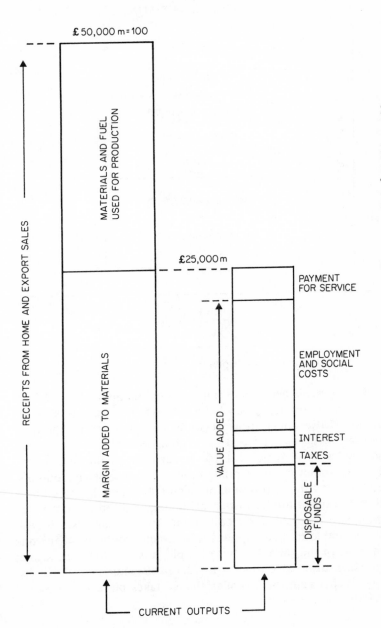

Figure 9.3 All manufacturing - pattern of costs in the United Kingdom, 1972 £50 000 = 100

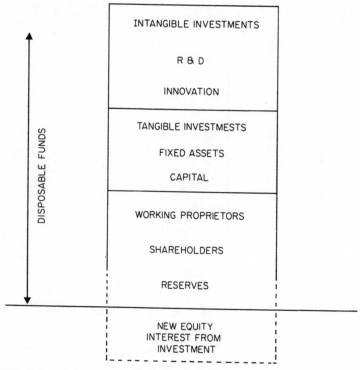

Figure 9.4 Distribution of disposable funds

Cox presents her data by adapting and modifying a technique first used by the European Industrial Research Managers' Association (EIRMA)(9). It expresses in percentage terms the relationship between sales and work done and the expenditure associated with activities directed at current and future business. Fig. 9.3 shows the appropriate heading for receipts and outputs, and Fig. 9.4 the distribution of disposable funds. In cases where R & D expenditure is written off and capitalized the latter diagram will need suitable modification. In both models it is relatively simple to study options that are created by changing one or more of the variables. For example, it is easy to see the extent to which disposable funds are reduced if material costs rise by, say, 50 per cent, or a similar change takes place in wages and salaries.

Cox used the above models as a basis for her analyses and obtained data which explained the observed differences in performance of the four United Kingdom and German industries. She demonstrated, in a convincing manner, the inevitability of a decrease in competitiveness and the consequential loss of jobs when a payroll budget grows at the expense of a satisfactory level of investment.

Higher pay can be one consequence of successful innovation, and a business should be so run that it can pay sufficiently high wages and salaries to secure and retain high quality staff without prejudice to its investment for the future. Cox's studies are too wide-ranging to be satisfactorily summarized, but it is important to note that her analysis showed that, for the four industrial sectors studied, the following factors were among those considered essential for competitive growth.

Revenue

1. The cost of materials and employment costs for current output should together be below 70 per cent of sales receipts.
2. The margin added to the materials purchased to be above the norm for the industry, and the rises in unit labour costs not to exceed rises in revenue.

Investment

3. The allocation of funds into investment devoted to future activities to be between 10 and 20 per cent of sales receipts.

Other noteworthy observations include:

1. The allocation of investment funds to research and innovation to be greater than to fixed assets.
2. New designs and extended commercial intelligence should have a greater priority than new equipment.
3. An industry must not spend more than 70 per cent of the value of its sales on materials and employment for current output, otherwise it will cease to be viable.
4. In times of economic stress the funding of staff for future output, and thus future jobs within the company, must not be

squeezed between the lower proceeds from sales and unions' demands for more pay on the shop floor.

5. A prudent norm for a payroll budget should ensure that current output does not take more than 31 per cent of the value of sales.

Cox believes that future activities can be financed from internally generated funds rather than by increasing debt and/or equity, a view which she supports by showing that funds from non-trading sources are only of the order of $\frac{1}{2}$ per cent of sales and can so be ignored without disturbing the analyses. Jones, on the other hand, believes the time has passed when United Kingdom industry can grow, or even survive, without external funds. Both authors conclude that United Kingdom industry needs better designed products of higher quality, but whereas Jones argues for lower priced products, Cox states that a persistent United Kingdom weakness has been its failure to move up-market and, as a consequence, has offered neither the high quality nor the excellent service needed to increase disposable funds.

Lorenz(10) has also attributed United Kingdom industry's loss of international competitiveness to a failure to invest effectively in the design of high quality, up-to-date products and processes. He cites the largest electronics manufacturer in West Germany which increased profit but cut dividends in order to provide funds for investment in future activities, and contrasts this with the United Kingdom industry which, when profits slump, nevertheless maintains dividends by cutting expenditure on research and development.

The added value model can help individual companies to anticipate better, the long-term effects of changes in cost patterns.

9.3 FINANCING RESEARCH AND DEVELOPMENT

A decision on the proportion of disposable funds which should be allocated to future activities does not necessarily indicate how much should be spent on research and development. It usually follows that the higher the calibre of a new product concept the lower are the likely associated costs of production and marketing, and that at the highest level of technology R & D expenditure is likely to be a major component of product costs. By contrast, at lower levels of technology a higher proportion of funds allocated to future activities will be spent on production plant and marketing. While the level of technology may be used as a rough comparative guide, it cannot indicate the actual magnitude of a research and development budget. It is therefore helpful to examine the experience of industrial groups of companies in a number of countries.

A treatment of the subject by Childs(11) is relevant since it differentiates between R & D aimed at current, and future, developments. The paper, again based upon the findings of an EIRMA working group, deals with eight industries and tabulates how 105 major European firms allocated their funds for research and development.

The EIRMA group divided R & D activities into the following categories:

A R & D for existing businesses
A1 Maintenance and minor improvements in manufacturing
A2 Major improvements and renewal in existing business, and for related diversification.
B New business for which neither the technology nor the marketing organization is available.
C Exploratory and basic research to generate knowledge leading to new business areas, and to keep abreast of developments in the basic technology of the business.

In Table 9.1, column I shows the classification of industrial groups, column II the number of companies which were surveyed in each group, and Column III the average total R & D expenditure as a percentage of company turnover. Table 9.2 shows the range and average percentage of R & D expenditure for categories A1, A2, B and C, from which it follows that new business categories B and C absorb approximately 20 per cent of the total.

This percentage is applied to column III of Table 9.1 and entered in column IV to show how expenditure on future activities varies among the seven industrial groups. The 1.8 per cent allocated to new business for group 1 is the highest for all categories; the average is 0.7.

Table 9.1   Classification of industrial product groups

Column I Industrial group	Column II Number of companies in group	Column III Average R & D expenditure as a percentage of turnover	Column IV Average total R & D expenditure allotted to new business as a percentage of turnover
1. (a) Aerospace (b) Electrical/electronic engineering (c) Instrument engineering	24	7.1	1.8
2. (a) Bulk commodity chemicals, plastics, synthetic fibres	15	3.4	0.9
(b) Speciality chemicals, pharmaceuticals, photo printing ink	16	5.3	1.3
(c) Fermentation, biotechnology			
3. (a) Mechanical engineering (b) Marine engineering/ shipbuilding (c) Civil engineering (d) Chemical engineering	9	2.1	0.5
4. (a) Steel (b) Non-ferrous metals (c) Building materials (d) Mining and quarrying	6	1.6	0.3
5. (a) Petroleum products and natural gas (b) Other energy	11	0.6	0.1
6. (a) Automobile engineering	9	2.5	0.5
7. (a) Glass pottery and china (b) Paper, and paper products (c) Timber and furniture (d) Rubber and plastic products (e) Natural textile fibres, textiles, footwear, leather (f) Food, drink and tobacco (g) Household products, detergents and toiletries	15	1.4	0.3

Table 9.2   Average percent of R & D funds spent on current and future
activities

Category	Percentage R & D expenditure by category	
	range	average
A1	21–44	34
A2	40–61	47
B	7–18	12
C	3–13	7

EIRMA's data can be compared with the results of the above value added analyses by making two estimates: the proportion of turnover invested in new business, and the percentage of the sum so invested allocated to research and engineering. Cox's recommendation for the former is from 10 to 20 per cent, and Jones' estimate for the latter ranges from 15 to 30 per cent. A figure of 15 per cent seems appropriate for the investment in new business on the grounds that Cox's work indicates that tangible assets should absorb less than intangible assets. Since the concern is with medium technologies 20 per cent seems an appropriate choice from the range given by Jones. These figures together indicate that expenditure on engineering and research directed towards new business should be 3 per cent of turnover, and this is over quadruple the average of 0.7 per cent of turnover which European firms were spending during the early 1980s (Table 9.1, column IV). Opinions vary on which activities should be included under the umbrella of R & D, and may well explain the wide range shown in Table 9.1 within each industry. Nevertheless, if the United Kingdom, and much of European industry is to regain competitiveness expenditure on research and engineering should be increased several-fold.

9.4 SELECTING IDEAS FOR DEVELOPMENT

Once a research development budget is agreed, a decision will

be required on the number and types of new products which require to be developed. If a search for ideas is carried out on a
continuous basis (see Chapter 5 page 105) a large number of possible concepts should be registered, and a small number should
merit further attention. An outstanding idea has two characteristics: its potential is easily recognized, and its evaluation will
not entail too many unknown factors. When uncertainties are
involved, the aim must be to seek additional data which will
enable them to be viewed as risks associated with an estimated
probability(12).

Selection of ideas proceeds in two stages. The first is qualitative, and ideas are sought which, generally, appear feasible. The
second phase is quantitative, and constitutes a more vigorous re-
examination of the earlier selection. Most qualitative methods involve drawing up checklists which comprise those factors which
are deemed necessary before an idea is accepted as a basis for
a project. Checklists which contain serious omissions can cause
considerable wastage of resources, and, to minimize the chance
of this happening, reference should be made to studies that have
been carried out on the identification of variables associated with
success.

In carrying out an investigation aimed at listing factors associated with successful marketing of new products, Cooper(13)
asked 177 firms to identify one failure and one success, and received data from 103. An unusually large number of variables was
examined, and the important conclusions showed that the specification for a new product must be drawn up more tightly than
that of a competitor, and should include reference to functional
performance, both product and operating costs, reliability, ser-
viceability, and compatibility. The risk of failure, furthermore,
is seen to be lessened if the product is a good fit with respect
to the abilities and training of the sales force, distribution and
market skills. Its design should facilitate satisfactory prototype
production and so facilitate satisfactory progress up to full-scale
production. This particular investigation by Cooper was based
on short-term considerations.

White's predictions for the determinants of success(14) concerned both quality and significance of the innovator's concept, and its embodiment in larger industrial devices or systems. He advocates that judgements be made on a number

of questions which include the extent to which (1.) a new product lifts fundamental constraints and introduces new ones; (2.) the act of embodiment necessitates further investigation of other components in the end product or whether it simply enhances the customer's innovation; (3.) existing business systems are strengthened or weakened by innovation or calling for new systems.

Table 9.3 Degree of innovation and organizational changes

Group	Degree of innovation	Suggested organizational change
1	p t m	existing organization
2	P t m	new engineering project team
3	p t M	new sales team
4	P t M	new product group
5	P T m	new product group
6	P T M	new venture company

*p = present product   P = new product
 t = present technology   T = new technology
 m = present market   M = new market

A paper by O'Leary(15) is, generally, in accord with White's views, but in it is a classification of particular relevance to the process of selection. The concept, shown in Table 9.3, suggests that thought should be given to assessing the degree to which a new idea affects the product, technology and market, and thus the organization. Recent work has indicated that this approach must be commonly adopted by managers, though perhaps subconsciously, since their perception of risk has been shown to be dependent upon the magnitude of the investment which, in turn, is seen as a function of the synergistic fit of the new idea with existing activities. Qualitative selection, based on this kind of intuitive decision, tends to reject long-term, high-risk projects which are necessary if a company is to enter new, dynamic markets, with new technologies. Horesh and Raz(17), in a study on the effect of doubt and uncertainty, have claimed that only one or two variables may affect an assessor's judgement on the viability of a new proposal. Among the most likely are the lifespan of the products, market size, and the estimated price of the new product.

Intuition, based on a wealth of experience and an unconscious weighing of many factors, may be very successful, but if the less experienced avoid what they perceive to be high-risk situations, and are content with an examination of only a few variables, there is clearly a role for quantitative techniques. These methods, however, still contain subjective elements in so far as they depend on cash-flow data, and require views to be taken of future project costs and profitability. Their big advantage lies in the need to identify factors which have to be explored in depth in order to assign probabilities, and the use of sensitivity analysis to minimize the number of variables likely to be significant. A simple, well-known, quantitative method is due to Disman(17) who uses an index number to indicate the worth of the project. It contains three elements: the gain or benefit which is usually some measure of expected profitability discounted to present values; the cost, again discounted, which normally includes R & D, production and marketing costs; and probabilities of both technical and commercial success. It is expressed as:-

$$\frac{P_G \times P_{TS} \times P_{CS}}{COSTS}$$

When P_G is the potential gain, P_{TS} is the probability of technical success, and P_{CS} the probability of commercial success.

The total cost of a project, and the ensuing benefits, are dependent upon the time taken to launch the new product, a quantitative selection method must begin with a view of the date at which the new product can be marketed. Should the actual greatly exceed the estimated time, a project is unlikely to be successful because of higher than expected costs and a failure to enter the market at a propitious time. Parker(18), in a retrospective analysis of 20 projects, equally divided between innovation work on new products and cost reduction work, showed the manner in which estimates varied with time. The results produced in Fig. 9.5, show that, in ascending order, the degree of reliability for the three factors is benefit, duration, and costs. This is not unexpected. Estimates of benefits extend furthest into the uncertain future, and delays caused, for example, by inadequate resources, may not necessarily increase costs.

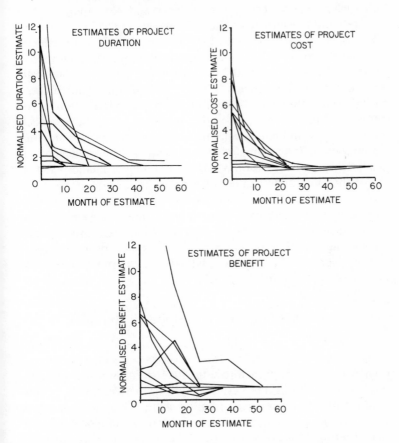

Figure 9.5 Normalized estimates of project durations, costs, and benefits

The two surprising features of the analysis were the large magnitude of the estimators' errors, and the discovery that other published work of a similar nature disclosed errors of comparable magnitude. It will be noted that in Fig. 9.5 all the innovations converged to unity half way through a project, and this suggests that major decisions should be deferred as late as possible consistent with starting a project on time. It is also clearly necessary to update estimates at regular intervals.

The tabulation of successive estimates were found to have a salutary influence on staff, who, subsequently, took care to review earlier experiences of project work and were thereby able to improve the accuracy of their estimates. The reiterative process was not developed on a methodological basis. Downsland, Das and Sharma(19) have described an iterative learning model which, when supplied with historical data, is able to simulate the decision-making process. Using the normal discounted cash-flow approach the model's main function was to indicate how resources should be allocated among projects during the appraisal stage, an early and late development phase, and the manufacturing operation. It was concluded that significant improvements in profits are possible if account is taken not only of financial status, but of the magnitude and complexity of the technology.

The difficulty of estimating probabilities has been one reason why companies have been slow to adopt the numerous sophisticated techniques described in the academic literature. Linear, integer, and dynamic programming models have been described, as have operational techniques that help the computation of cash-flows from probability inputs(20).

9.5 THE DISCOUNTED CASH-FLOW APPROACH

In a survey of 34 widely differing companies McRae(21) found that 75 per cent of them evaluated projects by using a discounted cash-flow (DCF) approach to estimate cash inflows and outflows. He believes it to be the best selection device because it is consistent with the acceptance of equity share values as an index of company success, and one which is strongly influenced by estimated future cash-flows from portfolios of activities in which the company is engaged. It is also in harmony with the company long-term goal which is to maximize share values and to minimize future cost of capital and so make cheap funds available for expansion.

The factors needed for discounted cash-flow calculation are cash inputs, cash outputs, and the cost of capital over the period of the project; all three should include only additional costs and revenue, and exclude any allocation of committed costs. Cash-flow should be stated net of tax and possible subsidies, and allowance must be made for risk and inflation.

Table 9.4 Cash-flow statement to calculate yield on product X

£'000		*a b c* Cash	*d e f* Net	Net cash flow discounted at	
Period	Cash inflow	outflow	cash flow	19 per cent	18 per cent
1	0	4	− 4	− 3.36	− 3.39
2	0	16	−16	−11.30	−11.49
3	12	24	−12	− 7.12	− 7.31
4	36	18	+18	8.96	9.29
5	32	18	+14	5.87	6.12
6	25	14	+11	3.87	4.07
7	20	10	+10	2.96	3.14
	125	104	+21		
Discounted net cash flow				− 0.12	+ 0.43

Note
A discount rate of 18 per cent reduces the net cash flow to a present value close to zero, this is therefore taken as the yield of product X.
The cost of capital to the company is calculated to be 8 per cent so, since 18 per cent is comfortably in excess of 8 per cent, the product is financially viable (McRae ignores risk and inflation)

Table 9.5 Calculating the cost of capital

	a	*d*	*c*	*d* *b x c*
	Source of funds	Amount £ million %	Net of tax cost, %	Weighted cost of fund
Loan	3	60	6	3.6
Rights issue of shares	1	20	12	2.4
Retained earnings	1	20	10	2.0
	5	100		
Average weighted cost of incremental capital -				8.00%

There are many ways of judging the cost of capital which must be invested in a project before cash begins to flow back into the company, and McRae recommends setting up the cost of capital as a hurdle (or discount rate) which must be exceeded before a product is accepted as financially viable. Table 9.4 and Table 9.5, taken from McRae's paper, are self-explanatory.

The discounted cash-flow method, even when used with reliable estimates of cash-flows, is widely criticized on the grounds that it filters out all but the low-risk mundane projects. McRae states that where blame exists it should not be directed at the method but of the way it is used. He directs particular attention to two factors: the premium for inflation is often crudely applied, and the predicted rate is added to the basic cost of capital rate, whereas it should be built into the model by adjusting individual cash-flows by anticipating future changes in inflation. Secondly, a decision should be taken as to whether the price of a new product varies with inflation or remains steady in money terms, thereby falling in real value and so becoming more price competitive. If high inflation makes the market less sensitive to price increases this may benefit the new product by allowing these price increases to absorb increasing costs and so the profit yield. Similarly, figures adopted for risks are too high. Some companies select a high premium to cover loss or failed products which is indefensible, while others often assume, incorrectly, that no portion of costs attributable to development, prototype production, or advertising, can be shared with, or switched to, current work or even other projects. In the period immediately prior to 1979 McRae found that the discount rates normally used ranged from 20 to 30 per cent, whereas a more realistic figure would have been 10 to 20 per cent. The probability that numerous acceptable ideas have been rejected by manufacturers, often with grave consequences to the firms concerned, reinforces the view that project evaluation should be regarded as a crucial stage of innovation. Ideally, information should be sought and refined until both qualitative and quantitative assessments are in accord. The hope is that inflation and interest rates will continue to reach low stable values, for only then can investments for the future be made with confidence.

9.6 SUMMARY

If a company is to generate sufficient funds to ensure continuity, production operations should be efficient and free from the unexpected. An important aim of designers should be to maintain a product range as small as possible, and manufacturing plant should undergo a major redesign every five years.

For a well-run company committed to long-term growth the value added should, in the long run, be sufficient to meet the cost of both current and future activities. If there has been a long period of growth, or negative growth, a business may survive only by means of an infusion of cheap capital. Reference is made to a macroeconomic study of the growth of firms in four industrial sectors, from which it was concluded that allocation of disposable funds devoted to future activities should range between 10 per cent and 20 per cent of sales receipts.

An estimate is made of the proportion of disposable funds that should be spent by research and development divisions on future activities. This shows that European companies in 1980, on average, underspent by a factor of four.

Ideas for new products require pruning to match resources, and it is customary to make an initial choice on qualitative grounds and to follow this with a vigorous quantitative examination.

Quantitative methods are briefly described but, as their success is seen to be critically dependent upon the accurate forecasting of development time, production costs, and future sales, the adoption of a reiterative learning model is recommended. The discounted cash-flow approach is the most commonly used technique for evaluating products, but care is needed when using it under conditions of high inflation.

9.7 REFERENCES

1. Barker, A. (1978). 'Some sacred cows to the slaughter', *Machinery and Production Engineering*, Editorial.
2. Small, B. (1983). 'Factories of the future—what will they look like', *The Production Engineer*, **62**, No.11, 30–1.
3. Jones, F. E. (1978). 'Our manufacturing industry—the missing £100,000 million', *National Westminster Bank Quarterly Review*, **May**, 8–17.

4. Jones, F. E. (1976). 'Pay, productivity and physicists', *Physics Bulletin*, **December**, 550–1.

5. Jones, F. E. (1977). *The Financial Environment*, Conference Economic Revival and Industrial Innovation, Presented by CBI, CEI, and IP & I.

6. Cox, J. G. (1977). 'Planning for technological innovation', Part I, Investment in technology. *Long Range Planning*, **10**, **December**, 40–4.

7. Cox, J. G. (1978). 'Planning for technological innovation: Part II—Investment in technological change in three major industries', *Long Range Planning*, **11**, **August**, 70–6.

8. Cox, J. G. and Kriegbaum, H. (1980). *Growth, Innovation and Employment: An Anglo-German Comparison*, Anglo-German Foundation for the Study of Industrial Society, London, 1–77.

9. Anon. (1974). *The Allocation of Research Resources*, European Industrial Research Management Association, Working Group No.12.

10. Lorenz, C. (1979). *Investing in Success: How to Profit from Design and Innovation*, Anglo-German Foundation for the Study of Industrial Society, London, 1–26.

11. Childs, A. F. (1983). *An Appropriate Level of R & D Spending for the UK*, R & D Society Symposium on The Survival of Industrial Research and Development. 15–30.

12. Cooper, R. G. (1981). 'The components of risk in new product development, Project New Prod., R & D Management, **11**, 2, 47–54.

13. Cooper, R. G. (1979). 'Identifying Industrial New Product Success': Project New Prod., Industrial Marketing Management, **8**, 136–144.

14. White, G. R. (1978). *Management Criteria for Effective Innovation*, Innovation Technology Review. Massachusetts Institute of Technology, Cambridge, Mass., 21–9.

15. O'Leary, R. (1978). 'Clues for product innovators', *Management Decision*, **16**, No.4, 195–207.

16. Horesh, R. and Raz, B. (1982). 'Technological aspects of project selection', *R & D Management*, **12**, No.3, 133–40.

17. Disman, S. (1962). Selecting R & D projects for profit', *Chemical Engineering*, No. 26. 24 December. 87–90.

18. Parker, R. C. (1974). *R & D Evaluation and Selection*, Institution of Mechanical Engineers Conference Publication 140, 12–19.

19. Dowsland, W. B., Das, D. and Sharma, K. G. (1981). 'Project selection using a learning process', *Chartered Mechanical Engineer*, **28**, No.2, 59–63.

20. Pessemier, E. A. (1977). *Product Management: Strategy and Organization*, John Wiley & Sons, New York.

21. McRae, T. W. (1979). 'Financing product innovation', *Engineering*, Issue January to June inclusive, 46–7, 135–6, 323–5, 446–7, 602–3, 752–4.

10

The Role of Government and Other Agencies

10.1 INTERACTIONS BETWEEN GOVERNMENT AND BUSINESSES

Industrialists frequently complain that the cumulative effect of government policies to aid business growth has been to hinder rather than to help(1,2). This view has arisen at least in part because government policies are so frequently amended, or even reversed, that companies operate within an atmosphere of great uncertainty. Parliamentarians are either not fully aware of the difficulties they create, or are overconcerned with short-term, political tactics. Every year much new legislation is passed which inevitably distracts companies from their planned paths and deflects both executives and managers from their main concern which is running a business. Policy reversals have been numerous; nationalization followed by denationalization, renationalization, and privatization, a refusal to join the European Community followed by a later demand for entry; a wage and salary freeze followed by a wage and salary explosion, and later by a voluntary restraint and social contract.

Direct intervention into industry has similarly been variable as to its results, and much of it ill-conceived. The encouragement given to industry to merge activities into even larger units was followed by a policy aimed at stimulating the formation of smaller business units. Numerous other contradictions could be cited, thus emphasizing the wisdom of the many pleas(3) for less intervention by government into industry.

The list of legislative changes is long and reflects swings between the two major political parties. Obstacles caused by policies not specifically aimed at industry are no less important, for as we saw earlier, sensible planning of investment is difficult when both inflation and interest rates fluctuate wildly. Investment policies have had to cope with an inflation rate which dropped to $4\frac{1}{2}$ per cent in 1971 and climbed to 13 per cent within 24 months, ranged from 14 per cent to 6 per cent in 1977, peaked at 17 per cent in 1979, varied between 5 per cent and $9\frac{1}{2}$ per cent in 1982, and dropped to a low of 3.8 per cent rate in 1983.

Changes in those government bodies concerned with research and development policies have been no less distracting. In 1964 the Department of Scientific and Industrial Research (DSIR) and the Advisory Council for Scientific Policy (ACSP) were abolished and five new bodies created: The Ministry of Technology (Min.Tech), a new advisory committee named the Council for Scientific Policy (CSP), a Ministry of Education and Science, and two Research Councils. A complete review of governmental research and development was later carried out by Lord Rothschild who recommended major changes in support of his customer-contractor principle (see page 230). Accordingly, in 1970, Min.Tech was replaced by a Department of Trade and Industry (DTI), CSP was abolished and a new Advisory Board of the Research Councils (ABRC) was set up. In 1976 an Advisory Council for Applied Research and Development (ACARD) was formed to complement the functions of ABRC. By 1983 the Department of Trade and Industry had been re-formed from the Department of Industry, and the Board of Trade. The perceived advantage of this latter reversion to an earlier arrangement is the creation of more effective links with matters affecting overseas markets.

Changes in policy which are intended to be beneficial should be neither resisted nor unduly criticized, but to introduce reorganization of the above magnitude at five-yearly intervals confuses both scientists and technologists, in government and in industry. Ronayne(4) has examined the above, and other, changes in detail and concludes that, despite the formation of three high level co-ordinating bodies, there is but scant evidence that the £3.5 billion budget for government research (including Ministry of Defence) is being satisfactorily monitored. He remarks that coordination is intradepartmental rather than overall.

Government schemes for encouraging new businesses are currently attracting most attention but, with the possible exception of enterprize zones, they have not been established long enough to justify comment. The first zone started operating in Swansea in 1981, and a 1983 report commissioned by the government stated that by May 1981, the first eleven zones had produced 2,900 jobs each at a cost of £20,000 per job (5), a figure that was later challenged and estimated to exceed £40,000 per job(6). Only half of the 300 companies, and 40 per cent of jobs, were said to be new, and the remainder were transfers which had moved locally. Despite these discouraging initial results 13 more zones were created in 1984. Hall observed that because the scheme is geared to savings on capital costs it tends to attract land-intensive rather than labour-intensive activities.

Despite many policy changes within the Department of Industry, the Government research establishments and industrial research associations have only been deflected by policy changes to a minor degree, and, indeed, have displayed considerable consistency in the excellent work which they carry out for the industries they serve. The comparative success of research establishments is attributable to close familiarity with the needs of their associated industries. This process was helped by the Rothschild customer–contractor principle, proposed in 1971, which recommended that applied research and development be carried out by an expert termed 'the contractor', for a customer who is prepared to say what he wants and for what he is prepared to pay. This encourages government research and development staff to place a practical evaluation on the results of their work. A dialogue between contractors and customers should, ideally, cover the widest possible area in which both parties are knowledgeable.

The types of assistance offered by the Department of Trade and Industry, which are particularly relevant to growth, include selective financial assistance for projects, grants towards research and development projects, advice or provision of financial assistance towards consultancy costs, assistance in technical appraisal, and funding of preproduction runs. The Department offers over a hundred diverse, separate, technical services (many of which can be financed by more than one source), and it is impractical to discuss here even the outline of the various services.

Unfortunately, several establishments may duplicate activities, and, despite the extensive publicity that is given to the many services, it is rare to find companies which use more than two or three, or are aware of the totality of help available. Few businessmen have sufficient time to absorb and understand the complications of the large government departments, and even fewer are versed in the intricacies needed to obtain information from European Economic Community bodies. Despite the chaotic environment which surrounds businesses, the survival of so many suggests that they have the resilience, will, and capability to prosper once they are allowed to operate with reasonable freedom. The basic need is for a supportive infrastructure which, *inter alia*, provides low price energy, a modern and acceptable transport system, and a foreign policy which does not disadvantage United Kingdom exports.

10.2 THE ROLE OF GOVERNMENT IN STIMULATING THE ADOPTION OF NEW TECHNOLOGY

Technology and businesses were brought closer together as a consequence of a National Economic Development Council (NEDC) meeting which was held at Chequers in 1975 and sought to emphasize the creation of wealth rather than its distribution. This led to the establishment of 18 economic development committees (EDC) and 32 sector working parties (SWps) which bring together representatives of specific sectors of industry, representatives of government departments, and trade union officials. These bodies perform a valuable service for their particular industrial sectors but services for general application come within the ambit of the Department of Trade and Industry.

Most Department of Trade and Industry schemes are concerned with transferring the results of applied science and showing how technologically advanced equipment can be adopted by a range of industries and services. The main beneficiaries are the new companies and the mature, large companies which operate at the high levels of technology. Little or no help is given to show how smaller companies, many of which are subsidiaries of large groups and operate at medium levels of technology, can innovate within their existing human and material resources. This is an important omission because it is this sector of industry which has the greatest potential for growth and for creating employment(7).

A body which has considerable potential for raising the technological level at which United Kingdom businesses operate is the British Technology Group (BTG). This body was formed in 1981 as the result of a merger of two public corporations: the National Enterprise Board and the National Research Development Council. The Group was designed to bridge the gap between the academic world and the world of businesses, and exploits technology derived from United Kingdom public sector sources and other public bodies. It assumes responsibility for protecting and licensing inventions from these sources, provides funds for development that may be required, seeks licences and negotiates licence agreements. BTG can provide funds for technical innovation and, depending on the circumstances, can provide finance either in the form of share capital or project finance. Small businesses are served by a Small Companies Division, and financial help is offered for small companies with growth potential. In 1983 the Group had a portfolio of 1700 UK patents and patent applications, and 600 licences in the United Kingdom and overseas.

Companies which were visited during the innovation studies described above (page xviii) were asked to list sources of external help which they had found useful within the last five years. The most widely used, and successful, was the DOI Manufacturing Advisory Service, and this can be attributed to the role played by a panel of dedicated, retired industrialists who are knowledgeable on industrial, technological, and commercial practices. They visit and inspect factories and advise how manufacturing processes can be improved. Indeed, the general impression gained from studies conducted in the many companies, was that the effectiveness of an assistance scheme depends upon it being active rather than passive.

From discussions on a need to encourage active interchange between those who offer and those who receive service have sprung many ideas, one of which is Harwell's initiative whereby groups of companies, which have expressed interest in one of their technological developments, are invited to share sponsorship of a research and development study. By this means individual companies only bear part of the risk associated with a high technological project.

A second example concerns the 'Teaching Company Scheme' which is jointly supported by the Science and Engineering Research Council (SERC) and the DTI. A partnership is formed between an academic institute and a manufacturing company with the aim of achieving a substantial improvement in production processes and, hence, in sales. High-class graduates are recruited, in consultation with the company, for two-year appointments as teaching company associates. The intent is that a fruitful liaison will develop between industrial and academic staff.

Hatfield Polytechnic has introduced a variant of this scheme— the Small Manufacturers' Industrial Development Association (SMIDA)—in which a committee, representing a number of small businesses, and chaired by the managing director of one of the companies, agrees to share the services of a science or engineering postgraduate. This arrangement prevents disillusionment which graduates of the Science Research Council 'Teaching Scheme' may sometimes suffer when smaller businesses become engulfed in firefighting and so are unable to spend sufficient time with the teaching company associates.

This principle of a group of small companies sharing services could, with advantage, be widely adopted, since there are many essential activities, as shown in the administration of overseas sales and preparation of transport documents, which do not require full-time staff.

10.3 OTHER WAYS IN WHICH GOVERNMENT CAN STIMULATE INNOVATION

From personal experience in industry, it is recommended that, where possible and appropriate, staff should be groomed for research and development directorships by undertaking an overseas technical mission of some months duration. A task should be chosen with a view to inculcating an awareness of market needs, emphasizing the importance of planning, finishing a programme on time, and demonstrating the interactions between science and technology. This principle of widening the experience of research and development staff is the basis of four suggestions whereby government initiative could stimulate innovation. The first involves the sponsorship of overseas market studies, the second the hosting of industrial teams in research establishments, the third the setting up of a central information centre, and the fourth encouragement of cooperation between producers and users by means of a wider use of procurement.

Turning now, to the first proposal, it is suggested that industrial manufacturing performance could be improved by the sponsorship of small groups, representing government laboratories,
higher educational establishments, and industry, for the purpose
of undertaking market surveys of high technological context in
overseas countries. Members of a group should consist of employees representing a wide range of responsibilities. The tasks
could take many forms, but, in view of the country's imperative
need to increase exports, a start could be centred on the Overseas Technical Information Unit of the Department of Trade and
Industry, and operate in conjunction with the technology coun-
sellors stationed at the Paris, Bonn, Moscow, Tokyo, and Wash-
ington embassies. This scheme could be regarded as a means of
complementing the British Overseas Trade Board (BOTB) Out-
ward Mission Scheme which encourages British exporters to visit
overseas markets by offering financial assistance to groups of exporters, provided they are sponsored by an appropriate Trade
Association or Chamber of Commerce. A valuable liaison function could result by working also in close harmony with the Export Intelligence Service of the BOTB and the Technical Help
to Exporters (THE), a service of the British Standards Institution which helps exporters to comply with standards and technical specifications for admission of products to overseas markets.
The emphasis would be on cross-disciplinary aspects, marketing,
economics, and technology, and the group should be recruited
from middle management.

In Chapter 4 it was stated that the effective transfer of technology is most fruitful when both those supplying and using information work in close proximity. It is suggested, therefore, that
when a company contracts to use a new, and possibly complex,
concept from a body such as a Government Research Establishment, or Industrial Research Association, the latter should agree
to host a small team from the company for the initial phases of
the development. Should unexpected directions problems arise,
staff with appropriate knowledge from other laboratories might
also be invited to join the team on a temporary basis. This arrangement would continue only until the company decided that
it had gained sufficient knowledge and experience to continue on
its own. A nationwide facility of this kind would be likely to be
of increasing value on two counts: it would lessen the rejection of

exploration of new technologies, by industry, on grounds of cost, and would stimulate the search for interdisciplinary solutions.

The earlier innovation studies clearly showed that few small and medium-size businesses know where to obtain information relative to their needs. It was observed that grave consequences frequently arose from inadequate information. The cost of products at their preproduction stage was not infrequently considerably above target, and this would not have been so had designers, or purchasing departments, located all possible suppliers of component parts. Instances were noted of dramatic cost reductions attained through chance knowledge of an acceptable alternative material. Again many companies sought ideas for new products in ignorance of relevant information contained on computerized databanks and in relevant publications. One striking illustration of the need for advice on technical alternatives was a statement made in 1981 that United Kingdom designers had available no less than 50 000 models of sensors comprising 5000 different types which could be grouped into 175 classes of basic parameters.

The difficulties of finding information can be partly ascribed to a plethora of sources(8) whose services overlap and confuse. The requirement is for a single initial 'enquiry point' to be sufficiently well known to enable the enquirer to be guided into the information network.

A universal and well-publicized 'first point of contact' could act as an 'information broker/catalyst', be simple, rapid, and yet comprehensive. It could deal with all possible questions, from the best source of finance for a particular purpose, to the procurement of nuts and bolts. Most enquiries would be dealt with by telephone but staff should be sufficiently knowledgeable to refer callers to appropriate organizations.

The centre would need a special inaugural publicity campaign, preferably with the help of television, aimed at those who could benefit most from it. It should be included in the telephone information services and detailed in the telephone directories' green pages. One solution would be to graft the concept on to the Department of Trade and Industry's Technical Advisory Point (TAP).

An objection that areas remote from London would not use the service could be overcome by cooperation with regional bodies, such as Chambers of Commerce. The Council for Small

Industries in Rural Areas (CoSIRA) might also act in a similar capacity, for it already offers comprehensive business management and technical advisory services, and can provide loans to small businesses in the depopulated areas in each county.

A referral service for manufacturing industries has operated in Germany and the Netherlands for over ten years, and over 50 per cent of small and medium-sized, and at least 20 per cent of large firms, claimed to have derived significant benefits. European experience has shown that there is an urgent need to help small to medium-sized companies embrace new technologies and so avoid the dangers that arise when existing products become obsolescent.

The stimulation of innovation through procurement is widely practised in Canada, Germany, and Sweden and is beginning to play a more significant role in the United Kingdom. For process innovation, procurement could enable manufacturers to introduce modern economic processes by stimulating demand for large quantity production. For newly innovated products, procurement should encourage early entry into the market, and so enable the supplier to launch his product ahead of possible overseas competitors. For highly sophisticated high technology products, government purchase is frequently a precondition for success and indeed, as illustrated in the case of the aircraft industry, purchasing agencies have become international.

Following successful purchasing policies in the military field, the United States set up, in 1973, an Experimental Technical Incentives Programme and attempted thereby to stimulate the development of a low cost blood analyser, a solar electric power plant, and a low noise air conditioning unit for households. This exercise was not successful. Dr Prakke(9) of the Netherlands' TNO organization suggested in 1979 that government purchasing policy should aim to encourage innovation in the public sector. He listed as suitable topics care of the elderly, medical technology, fire protection, block heating, and port facilities.

Procurement by government and local authorities should be reviewed carefully, and should avoid the likely ill-effects associated with too much intervention by limiting its actions to the establishment of an initial market launch. The prime purpose should be to encourage the closest cooperation between the producer and user, again by inviting a team representing both parties to share an appropriate laboratory or by testing facilities during crucial stages of development.

10.4 STANDARDS, SPECIFICATIONS, QUALITY ASSURANCE, AND PRODUCT LIABILITY

In so far that regulations, requirement, standards, specifications and product liability can affect the competitiveness of manufactured products, and involve international collaboration, they must be an important concern of governments. Exporters from the United Kingdom face obstacles due to these causes more frequently than do importers into this country. This situation does not necessarily arise because our competitors purposely erect barriers against United Kingdom manufacturers but rather because our competitors are more concerned with environmental matters and safety, whereas few United Kingdom regulations are mandatory.

Conditions which have to be met by the manufacturers of pressure vessels provides an interesting illustration; 16 countries have national codes for the design and construction of boilers and pressure vessels, and, of these, 15 give them the force of law either by direct legislation or ministerial decree. The United Kingdom has no form of legal requirement beyond requiring the user to ensure the products are 'of good construction, sound materials, adequate strength and free from patent defect' — not a precise definition. Against this background the Institution of Mechanical Engineers set up its Pressure Vessel Quality Assurance Board, which gives its certificates of approval to a number of independent inspectorates. This action clearly improves the competitiveness of British manufacturers since they will no longer have to wait for countries to send their own representative to ensure satisfactory compliance with their published requirements.

BSI standards are rarely mandatory, although their status may be enhanced by the use to which these are put by official bodies(10). It must be recognized that standards can have many functions and are drawn up in different ways. They should reflect a consensus of good practice and current knowledge, but yet be sufficiently flexible to give the innovator scope for originality.

The EEC regulations are important for all manufacturers, and while those concerned with safety may restrict design freedom, the majority should not. Optional harmonization directives are less restrictive than total directive regulations but even the latter must permit the most competent manufacturer to gain a major market share. To unify international standards is clearly a task which will always consume an acceptable amount of committee time and will therefore be costly. Recognition of standards is the preferred alternative.

The general question of accepting specifications is a critical one and a manufacturer may meet a number which refer to one class of product. They may include a US state regulation, an EEC harmonization regulation and a company acceptance test, and because they differ it may be necessary to manufacture several modifications of a base product. Considerable cost may be incurred when customers' specification tests fail to correlate with service performance, because substandard products may have to be produced for more than one market. Again, the need to meet several different international safety regulations can mean a choice between manufacturing several products to cover all markets, or making one product which is over-engineered for many of its applications.

The relationship between product liability and product specification is a matter of great importance, and serious consequences could follow if draft enactments are drawn up by bodies that are unfamiliar with the manufacturers' engineering capability and product usage. Manufacturing companies should be kept fully informed of the progress of developments in product liability. This question of keeping industry informed is vital in many other allied fields. The work involved in completing a dossier covering even one product and one country can be considerable, and pains should be taken, when drawing up regulations, to avoid unnecessary complexities. The process of collecting and imparting information in these various countries will constitute one of the measures referred to in Chapter 4.

Quality assurance is a subject which can be linked with the function of procurement. Products which are supplied, in the first instance, for a government department and receive their Seal of Quality assurance, are in a strong position when they extend sales into commercial channels; for example, British Specification 9000, which was drawn up primarily for NATO defence contracts, is also recognized by the Defence Quality Assurance Scheme.

The problem for the future is to establish collaboration between standards and certification authorities to ensure that regulations do not interfere with international trade, and to recognize that, while specifications and quality assurance can encourage sound design and quality, they must not restrict progress.

10.5 SUMMARY

Government policies which directly or indirectly affect business have undergone frequent reversals and the cumulative effect has been to create uncertainty, and so hinder rather than help businesses. The form and responsibility of government bodies concerned with research and development policies have undergone radical change on three occasions within the last 14 years, and coordination of government researches is believed to be unsatisfactory. The government's research establishments and research associations have continued to carry out excellent work despite the many changes in policies.

There are numerous schemes which have been designed to introduce technology into business. Several examples of better, and more successful, schemes are described and all share the virtue of being more active than passive. Most services are aimed at either high technology manufacturing industries or new businesses. Little is done to assist the large number of small companies which operate at medium levels of technology and are experienced, have unused resources, and consequently large unresolved potential for growth.

Suggestions are given for improving industrial performance based on a belief that it would be advantageous were small groups drawn from industry, government, and academia, to be asked to undertake market surveys of high technological products in overseas countries, or to develop products from concepts originating either in universities or government laboratories.

Data relevant to development work and design is difficult to obtain owing to the large number of overlapping information services, and it is recommended that a universal information/catalyst service be established which should be simple, rapid, and yet comprehensive.

Procurement policies should be operated by both government and local authorities. They should promote the closest cooperation between producers and users, should help manufacturers to install efficient processes by promise of substantial initial orders, and should encourage innovators to enter markets promptly.

Regulations, standards, specifications, quality assurance, and product liability pose difficult problems which may interfere with international trade. Small businesses are unlikely to employ staff capable of dealing with the many intricacies involved, and simple guidance should be available from national bodies or other agencies.

10.6 REFERENCES

1. Jewkes, J. (1977.) *Delusions of Dominance,* Hobart Paper. The Institution of Economic Affairs, 1–64.
2. McIntosh, R. (1978). 'Britain's industrial goals', *Production Engineering,* **57**, No.2, 36–40.
3. Grierson, R. H. (1978). 'The mirage of the state's entrepreneurial role', *Journal of the Royal Society of Arts,* **126**, No.5263, 400–6.
4. Ronayne, J. (1984). *Science Policy in the United Kingdom.* Edward Arnold (Publishers), London.
5. Hall, P. (1984). 'Facts, fantasy and lost opportunities'. *Financial Times,* **22 February**, 15.
6. Holloway, A. C. (1984). *Financial Times,* **14 March**, p.13. Letters to the Editor.
7. Parker, R. C. (1983). 'The contribution by smaller manufacturing companies to employment'. *BIM Management Review & Digest,* **10**, No.2, 7–9.
8. Department of Trade and Industry (1983). 'Guide to industrial support'. *British Business,* No.2, **January**.
9. Prakke, F. (1979). *Government Procurement Policies and Their Effects on Industrial Innovation—International Experience,* **February**, 12th International TNO Conference.
10. Department of Trade. (1982). *Standards, Quality and International Competitiveness* (1982). HMSO, 1–23.

INDEX